D0209364

Structure and Spectra
of Molecules

Structure and Spectra of Molecules

W. G. RICHARDS

Physical Chemistry Laboratory
Oxford University

P. R. SCOTT

Charterhouse
Godalming

JOHN WILEY & SONS

Chichester · New York · Brisbane · Toronto · Singapore

Copyright © 1985 by John Wiley & Sons Ltd.

Library of Congress Cataloging in Publication Data:

Richards, W. G. (William Graham)
  Structure and spectra of molecules.

  Includes index.
  1. Molecular spectroscopy.   2. Molecular structure.
I. Scott, P. R.   II. Title.
QC454.M6R54   1985   541.2′2   84–15333

ISBN 0 471 90577 1 (cloth)
ISBN 0 471 90579 8 (paper)

British Library Cataloguing in Publication Data:

Richards, W. G.
  Structure & spectra of molecules.
  1. Molecular structure   2. Molecular spectra
  I. Title   II. Scott, P. R.
  539′.12   QD461

ISBN 0 471 90577 1 (cloth)
ISBN 0 471 90579 8 (paper)

Typeset by Activity Ltd., Salisbury, Wilts
Printed by Pitman Press Ltd., Bath, Avon

Contents

Preface

Two factors have prompted us to write this book. One is the popularity of our *Structure and Spectra of Atoms*, to which this book can be considered complementary. The other is dissatisfaction with existing texts. Unlike the atomic situation, there is no shortage of books on molecular spectroscopy, although most are rather difficult for the beginner. The almost universal defect is, however, that they deal with spectroscopy without molecular structure, frequently giving the impression that molecular spectroscopy is just a game for spectroscopists whose aim is to unravel complex spectra with little heed to more concrete goals such as the understanding of structure.

We have followed the pattern of the earlier book, starting with an essentially non-mathematical account of molecular structure before treating each range of the electromagnetic spectrum in separate chapters. The spectroscopic chapters themselves contain much of a structural nature and we have tried to avoid, even within the covers of this book, any separation of structure from spectra.

The treatment is condensed to enable a student to have an introductory grasp of the whole range of molecular spectroscopy before delving into more advanced texts for the minutiae with which each branch of spectroscopy abounds.

WGR
PRS

Chapter 1

Introduction

Physicists have been known to say that a diatomic molecule is an atom with one nucleus too many! There is some justification for this bizarre statement as far as spectroscopy is concerned. The energy levels of an atom which give rise to the characteristic line spectra of atoms arise from altering the energy of the atom by changing the energy of the orbiting electrons. The same thing can be done for a molecule but in addition to its electronic energy a molecule will have vibrational, rotational and translational energy. Fortunately these extra types of energy may be considered separately but they do increase the complexity of molecular spectra in comparison with atomic spectra. By way of compensation the spectra reveal full details of molecular structure—not just the energies of the various electrons but also the frequencies of vibration of bonds, the bond lengths and the angles between them.

A. Quantized energy levels

The experiments in the early part of this century by Planck, Bohr and Einstein showed that the energy of a microscopic system such as an atom or a molecule is quantized; only certain definite energies are allowed. These experiments provided the foundation for quantum mechanics which describes correctly the behaviour of small particles and predicts the allowed levels of simple species. In quantum mechanics the energy levels of a system can in principle be found by solving the Schrödinger equation.

The energy of an atom or molecule depends primarily on which orbitals in the system are occupied by electrons; this is shown schematically in Figure 1.1 along with the corresponding energy diagrams. The separation of these electronic energy levels is large, of the order of tens of kilojoules per mole.

The energy of an atom is defined fully by its electronic configuration, but a molecule may have other types of energy. A molecule does not have a constant bond length, r, as in Figure 1.1, but undergoes regular vibration like two weights connected by a spring (Figure 1.2). This vibrational energy is again quantized so that only discrete frequencies (which depend on the masses of the

2

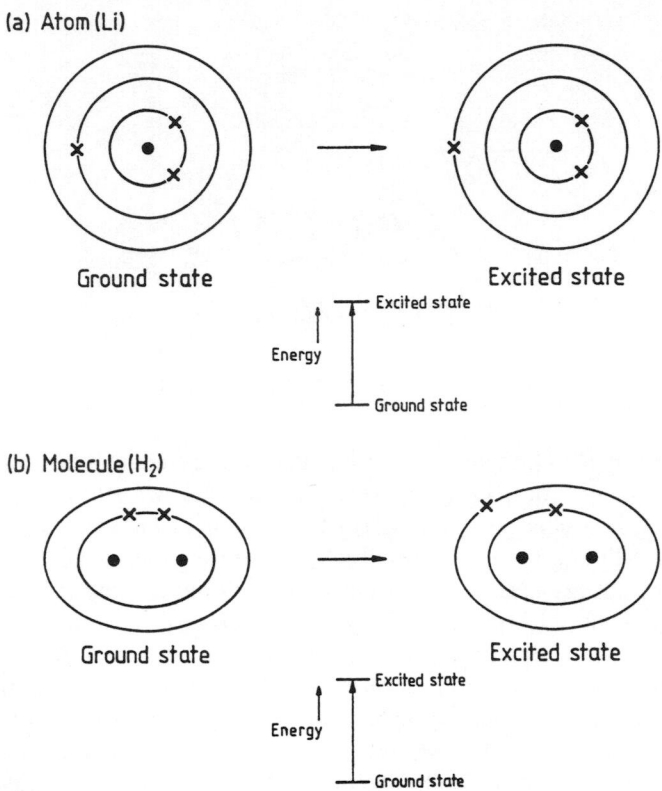

(a) Atom (Li)

Ground state   Excited state

Excited state

Energy

Ground state

(b) Molecule (H_2)

Ground state   Excited state

Excited state

Energy

Ground state

Figure 1.1 The excitation of electrons to give excited electronic states

(a) Diatomic molecule

(b) Triatomic molecule

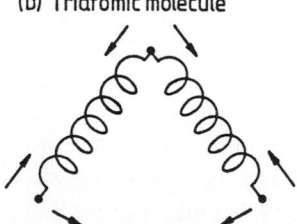

Figure 1.2 The vibrations of molecules with bonds treated as if they were 'springs'

atoms and the strength of the bond) are allowed. The separation of vibrational levels is less than that of electronic levels, being typically of the order of a few kilojoules per mole.

The rotational energy of a molecule as a whole is further quantized (Figure

(a) Diatomic molecule   (b) Triatomic molecule

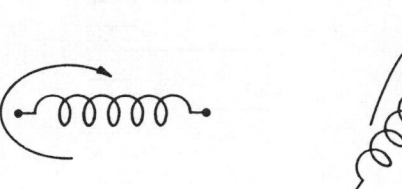

Figure 1.3 Rotations of molecules (diatomic molecules
have rotations about 2 axes and polyatomic molecules
about 3)

1.3). Rotational energy levels are even closer together with a separation of
some tens of joules per mole.

These three major forms of energy may be summarized in a single energy
level diagram as in Figure 1.4. Each electronic energy level has associated with
it a number of vibrational levels, each of which in turn has superimposed a set
of rotational levels.

Translational levels in principle add a further fine division of each rotational
level of the spectrum at levels, but although translation is from a theoretical
point of view quantized (from the motion of the particle in a box), in practice
the translational energy levels are so close together as to be virtually
continuous in energy. In spectroscopic terms they form a continuum.

The atomic nuclei in a molecule can sometimes have different energies
according to the orientations of their spin in a magnetic or electric field. These
are very closely separated in energy, closer even than rotational energy levels.
In many cases they may be neglected but do give rise to very important
applications in spin resonance spectroscopy.

B.   The energy of electromagnetic radiation

Electromagnetic radiation includes not only visible light but a wider spectrum
from X-rays and ultraviolet light on one side of the visible portion of the
spectrum, to infrared, microwave and radiofrequency radiation on the other.
All these forms are essentially the same save for their frequency, ν, or their
wavelength, λ. The frequency and wavelength are related, as all electromagnetic radiation has the same velocity, c, the speed of light. Therefore:

$$c = \nu \lambda$$

Thus high frequency light has a short wavelength while low frequency radiation
corresponds to a long wavelength.

The actual property which varies in a wave-like manner is in fact an electric
field with a magnetic field at right angles to it (hence the name electromagnetic
radiation) (see Figure 1.5). In many forms of spectroscopy we can ignore the
magnetic component and consider only the oscillating electric field, although

4

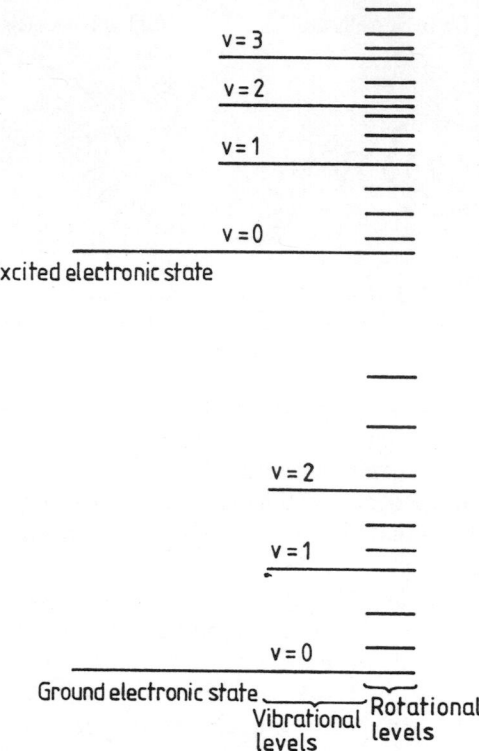

Figure 1.4 Schematic view of the manifold of energy levels: electronic, vibrational and rotational

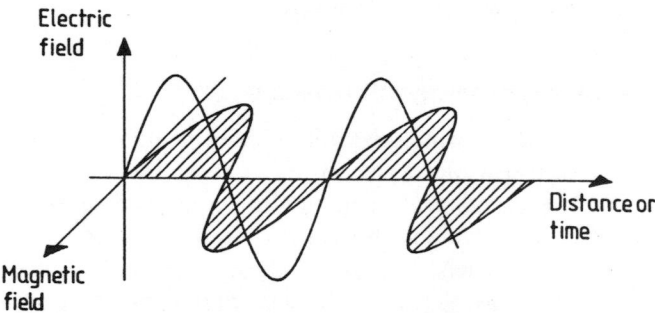

Figure 1.5 The perpendicular oscillating electric and magnetic fields of electromagnetic radiation

magnetic resonance experiments depend on the magnetic component of electromagnetic radiation.

Radiation also behaves as if it has the nature of particles or photons each with energy, E, depending on the frequency of the radiation, ν, given by Planck's relation:

$$E = h\nu,$$

where h is Planck's constant.

Altering the energy of a molecule from one level to another will involve the absorption of a photon if we are moving up the energy level diagrams in Figure 1.4, or the emission of a photon if the molecule moves from an energy level high on the diagram to the lower value. The frequency of the photon absorbed or emitted depends on the energy gap, ΔE, between the two energy levels

$$\nu = \frac{\Delta E}{h} = \frac{E_{initial} - E_{final}}{h}$$

because only photons whose energy matches ΔE precisely are absorbed in molecular spectroscopy; it is not generally possible for a molecule to absorb a photon with too much energy and emit the excess as a photon of lower energy.

Since the energy gaps depend on the type of energy being considered, different portions of the electromagnetic spectrum are involved in different molecular processes. Visible light and ultraviolet light excite electrons to higher orbitals; this excitation can subsequently produce chemical changes, and so light will fade clothes or cause suntans. Higher energy will even break up molecules. Infrared light will make molecules vibrate more, the extra energy being detectable as heat. Microwave radiation makes molecules rotate faster and is used in cooking as a way of putting energy into molecules. Radiofrequencies only have enough energy to reorientate atomic nuclei in magnetic fields and fortunately do not have any direct physiological effect. This important connection between the energy of photons and the type of molecular spectroscopy in which they are involved is summarized in Figure 1.6, which also serves to provide a conversion table between the various units used to measure energies.

As frequency and energy are directly proportional to each other, being related by Planck's constant, working spectroscopists are often very lax and talk of frequencies as if they were energies: a large frequency corresponds to a large energy. In the same way the reciprocal of the wavelength is also proportional to the energy $1/\lambda = E/hc$, and this is also commonly used as a measure of energy even though its units do not have the correct dimensions. The 'wavenumber' is the reciprocal of the wavelength in centimetres; it is written cm^{-1} and read as 'centimetres to the minus one'. The wavenumber scale is included in Figure 1.6.

The use of wavenumbers is not quite as curious as it appears at first sight since, as we shall see, spectroscopists actually measure wavelengths. To convert to frequencies or energies is in principle simple if we know the velocity of light, c. Although c is known quite accurately it is one of the less well determined of the fundamental constants. Hence if spectroscopists leave their energies in the wavenumber unit (cm^{-1}) then any subsequent

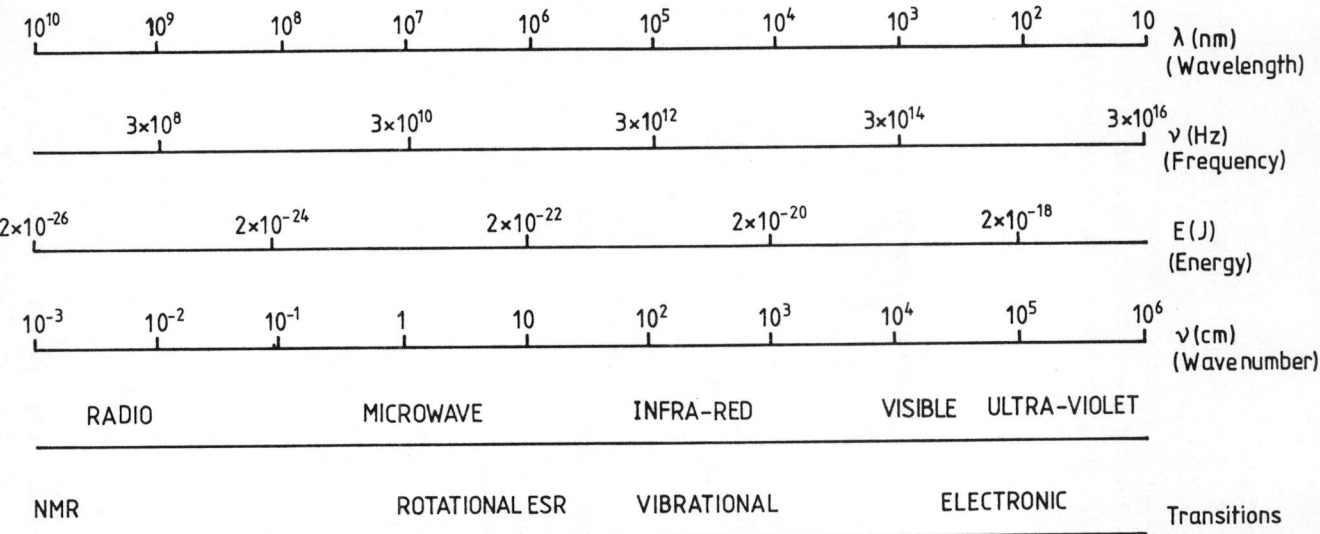

Figure 1.6 The electromagnetic spectrum

reevaluation of *c* will not alter their published work and the new constant can be used if it is necessary to convert to more conventional energy units.

C. Spectroscopy

For a molecule, even a diatomic molecule with only one bond to vibrate, the complete energy level diagram nonetheless contains a vast number of discretely allowed energy levels. If it were possible for a transition to occur between any pair of levels with the absorption or emission of a photon, then the spectrum would appear hopelessly complicated. There is, however, a major simplification in the form of selection rules. Not all transitions are possible. Only certain limited transitions are allowed, the rest being forbidden by the selection rules. The rules can be derived from quantum mechanics quite simply and usually depend on some symmetry restriction. They will be discussed in detail in later chapters when we consider the various energy regions of the molecular spectrum.

Experimentally there is a distinction between spectroscopy in absorption and in emission. In the former case continuous radiation which contains all frequencies is shone into a sample and the frequencies absorbed are detected. In emission, energy is pumped into the molecules, e.g. by heating or electrical discharge, to excite molecules to higher energy levels. The radiation emitted as molecules fall back to lower energies is then detected (Figure 1.7).

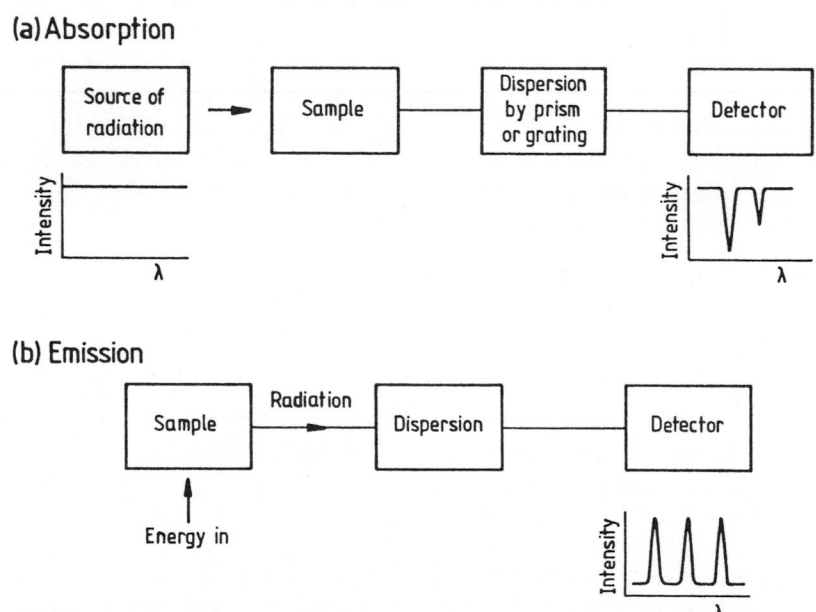

Figure 1.7 Block diagram of the experimental arrangements for absorption and emission spectroscopy

The continuous radiation may be dispersed into a spectrum using a prism or a diffraction grating and detection of the dispersed spectrum may be by a photographic plate or by using a photomultiplier and appropriate amplification to work a pen recorder so that the spectrum actually appears on a paper trace.

If spectroscopy is to be quantitative then it is of course necessary to convert the peaks on a paper trace or the lines on a photographic plate to wavelengths by comparing them with peaks produced by known standard wavelengths. These are themselves spectral lines whose wavelengths have been measured accurately using an interferometer and which are to be found in tables of standards. These standard spectral lines have ultimately been compared to the standard unit of length; this is no longer a bar of platinum-iridium alloy whose length is defined as 1 metre and is kept in Paris but is defined by reference to a line in an atomic spectrum (currently the wavelength of a line in the spectrum of the caesium atom).

D. The population of energy levels

Any molecule will have associated with it a vast number of possible energy levels, electronic, vibrational and rotational, with further hyperfine components if we place the molecule into a magnetic or electric field. If we have a large number of molecules they will not all occupy the same level. For a statistically significant number of molecules such as one has in a laboratory sample of gas, liquid or solution, the actual distribution of molecules among the possible energy levels on the energy diagram depends on the temperature of the sample.

Qualitatively, at low temperatures the molecules are to be found occupying the energy levels low down on the diagram and heating gives rise to increased populations in higher levels (see Figure 1.8). Quantitatively, for a given temperature, T, the population decreases exponentially as we go up the diagram. (This is expressed as the Boltzmann distribution law: the population level of energy, E, is proportional to $e^{-E/kT}$.)

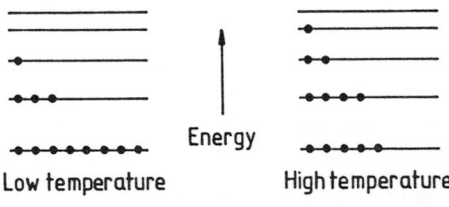

Figure 1.8 The variation of populations with temperature

In a spectroscopic experiment we will start with a given distribution of levels. When energy is absorbed higher levels become populated. If the molecules were then merely to sit in these newly occupied levels there would come a time when no further absorption was possible at a given wavelength since there would be no molecules in the lower level. The spectrum would then disappear and be said to be saturated. This does not happen in practice because molecules collide with each other and with the walls of the container. They can transfer energy as heat (vibrational or rotational energy) and restore the equilibrium population so that the spectrum does not become saturated. This process is called relaxation and does not complicate spectra save for the radiofrequency region where the energy levels are very close together. This point will be discussed when we consider nuclear magnetic resonance at the end of this book.

E.   Applications of spectroscopy

When studied in detail the spectrum of a molecule can be used to establish the structure of that molecule. This is the application stressed in this book. More routinely, spectroscopic measurements can be used to identify a substance and possibly determine its concentration. Spectroscopy of molecules is also the technique to use to answer questions about the local environment, such as on a catalyst or solubilized in a micelle, or to see how rapidly the molecule is undergoing some change, perhaps in reaction.

F.   Summary

Molecular spectroscopy involves transitions between the quantized energy levels of the molecule. Ultraviolet and visible radiation is associated with changes in electronic energy levels, infrared radiation with changes in vibrational energy and microwave radiation with changes in rotational energy.

Changes in rotational energy can take place with no other changes on the energy diagram. Vibrational transitions have associated rotational changes. Electronic transitions have simultaneous changes in vibration which again have associated rotational fine structure.

Thus the simplest logical way to approach molecular spectroscopy is to start with pure rotation, then to consider vibration-rotation transitions and finally electronic spectra. At each stage the knowledge of part of the energy level diagrams built up previously will still be relevant.

Before going directly to spectra, however, we should say something more about molecular structure since it must be emphasized that spectroscopy exists not as an amusement in itself but as a means to understand structure.

G.   Problems

1. Using $c = \nu\lambda$, with $c = 3.0 \times 10^8$ ms^{-1}, calculate the wavelength of a photon that has a frequency of 1.2×10^{15} Hz. What is the energy of the photon in joules per photon and in kilojoules per mole?
2. Calculate the wavelength of a photon of visible light with a frequency of 6.6×10^{14}Hz. What is the energy of the photon in joules per photon and what is the wave number?

3. The first excited electronic state of molecule lies at 200 kJ mol^{-1} above the ground state. Using the Boltzmann distribution law and the fact that at room temperature RT is roughly 2.5 kJ mol^{-1}, find the proportion of molecules in this excited state.

4. For CO the spacing of vibrational levels 0 and 1 is about 15 kJ mol^{-1}. What proportion of molecules are in the state with $v = 1$?

5. Rationalize the fact that fluorine (F_2) is colourless, chlorine (Cl_2) is green and bromine (Br_2) is red in terms of the energy required to excite electrons in the molecules.

Chapter 2
Chemical Bonding

Familiarity with atoms, their chemistry and their spectra has made us accept quite happily that the 'structures' or electron configurations of the ground states of some simple atoms are:

H : $1s$
Li : $1s^2\, 2s$
F : $1s^2\, 2s^2\, 2p^5$

If we excite the $2s$ electron in lithium to the $2p$ orbital then this corresponds to the excitation seen as the D lines in the spectra of alkali metals, as illustrated in Figure 2.1.

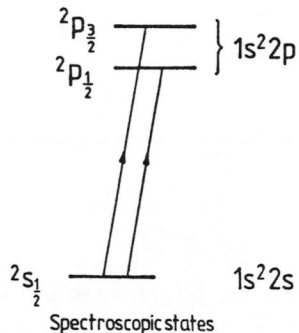

Spectroscopic states

Figure 2.1 The transitions involved in the D lines of the lithium spectrum

So commonplace is this habit of thinking of the properties of an atom in terms of its electronic structure that the true meaning of the notation which indicates that helium has a 'structure' $1s^2$ is forgotten. In this chapter we will extend the idea of electronic structure to diatomic molecules, to illustrate the principles of chemical bonding.

The common description of an electronic structure depends on a mathematical approximation known as the orbital approximation.

11

A. The orbital approximation

The Schrödinger equation for an atom or molecule may be written in shorthand form as

$$H\Psi = E\Psi,$$

where Ψ is the wavefunction for the system and E the energy. H is an operator which contains details of the interactions of all the particles in the system. It is not possible to solve the Schrödinger equation exactly for systems with more than one electron, and so the approximation is commonly made that the total wavefunction Ψ may be represented as a product of functions each giving the behaviour of an individual electron:

$$\Psi = \phi_1\phi_2\phi_3 \cdots \phi_n.$$

This is equivalent to assuming that each electron is influenced only by the average positions of the other electrons, and not by their instantaneous positions.

For a helium atom we can now write

$$\Psi = \phi_{1s}^{\alpha}(1)\ \phi_{1s}^{\beta}(2),$$

where ϕ_{1s} is a three-dimensional function often represented as in Figure 2.2. The value of ϕ^2 (or more correctly ϕ multiplied by its complex

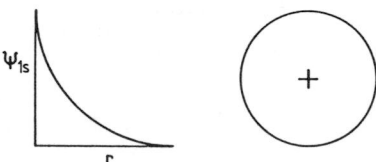

Figure 2.2 Representations of a $1s$ atomic orbital

conjugate, although there is no difference in this case) at any point gives the probability of finding an electron there, or can be thought of as measuring the electron cloud density. The superscripts α and β refer to the electron spin, which can take only two values, sometimes thought of as spin 'up' and spin 'down'. In the helium atom the two electrons have opposed spins, as they both occupy the same orbital.

The wavefunction we have written so far is not quite satisfactory, as according to the Pauli principle the wavefunction must change sign if two electrons are exchanged. This is a fundamental property of electrons, and can be satisfied if we rewrite the wavefunctions as

$$\Psi = \phi_{1s}^{\alpha}(1)\ \phi_{1s}^{\beta}(2) - \phi_{1s}^{\alpha}(2)\ \phi_{1s}^{\beta}(1).$$

If the labels (1) and (2) are now exchanged, then Ψ becomes $-\Psi$. In mathematical terms, the wavefunction is now the determinant of a matrix:

$$\psi = \begin{vmatrix} \phi_{1s}^{\alpha}(1) & \phi_{1s}^{\beta}(1) \\ \phi_{1s}^{\alpha}(2) & \phi_{1s}^{\beta}(2) \end{vmatrix}$$

If we consider an atom with more than two electrons, the wavefunction must still change sign on the exchange of any pair of electrons. The full form of the wavefunction is now more complex, as it involves a linear combination of more terms, but the correct combination is still obtained by taking the determinant of a matrix. Thus for a lithium atom,

$$\psi = \begin{vmatrix} 1s^{\alpha}(1) & 1s^{\beta}(1) & 2s^{\alpha}(1) \\ 1s^{\alpha}(2) & 1s^{\beta}(2) & 2s^{\alpha}(2) \\ 1s^{\alpha}(3) & 1s^{\beta}(3) & 2s^{\alpha}(3) \end{vmatrix}.$$

Writing out all six terms is now very tedious, and so the wavefunction of lithium is commonly written as $1s^2\,2s$, but this is only a convenient shorthand for a lengthy sum of terms.

One of the consequences of the determinantal form of wavefunctions is the familiar rule that no two electrons in an atom may have all four quantum numbers the same. If two electrons were absolutely identical, then two columns in the determinant would be identical, and the value of the determinant would then be zero.

From the point of view of the orbital approximation, there is no difference between atoms and molecules. In the case of the latter a molecular wavefunction may be expressed approximately as a product of one-electron functions, each with spin α or β, and the whole set of products involving all interchanges of electrons must be included to satisfy the Pauli principle.

Thus for some simple molecules we may, using the same shorthand, write their 'electronic structures' in the following way:

$$
\begin{aligned}
&\text{H}_2 & : \quad & 1\sigma_g^2 \\
&\text{Li}_2 & : \quad & 1\sigma_g^2\,1\sigma_u^2\,2\sigma_g^2 \\
&\text{CO} & : \quad & 1\sigma^2 2\sigma^2 3\sigma^2 4\sigma^2 1\pi^4 5\sigma^2
\end{aligned}
$$

Many aspects of the atomic notation carry over to the molecular situation. The running numbers $1, 2, 3$, etc., increase as the energy of the orbital goes up and they are like the atomic quantum number n. The symbols σ_g, σ_u, π, etc., give an indication of the symmetry of the one-electron functions for the molecule in a manner parallel to that coded as s, p, d, f, etc., for atoms. The superscripts reveal the number of electrons in the orbitals, and in the examples cited above all the electrons are paired with partners of opposite spin so that we have closed shell or rare gas-like stability. The overwhelming number of molecules dealt with by chemists do have closed-shell ground states, although a few molecules such as NO are free radicals which have a single unpaired electron.

For the theoretical chemist a major goal is to determine the nature of molecular wavefunction ψ in terms of molecular orbitals ϕ_i. The most common method of doing this is to break down the molecular orbitals further into expansions of other known functions. The qualitative aspects of this are important in understanding molecular structure and spectra, but before discussing the elements of this procedure there is a further approximation which is necessary before we can treat molecules by an extension of methods suitable for atoms.

B. The Born–Oppenheimer approximation

The added complicating feature possessed by molecules but not by atoms is that the energy of a molecule depends on the relative positions of its constituent atoms. Even without exciting electrons to higher orbitals, the energy of the H_2 molecule will vary as the H–H bond is stretched or compressed, and the energy of H_3 is not the same when the atoms are arranged in a straight line and when they are in an equilateral triangle.

The Born–Oppenheimer approximation allows us to handle this extra complication. The approximation treats nuclear and electronic motion as entirely independent; this is reasonable, as the velocity of nuclear motion is very much less than that of the much lighter electrons.

In practice this means that we consider a molecule with a given nuclear framework and calculate its electronic energy only. The nuclear–nuclear repulsion is then added on as a separate term, and is calculated as a simple sum of coulomb repulsion energies. We may then repeat the process for a different set of coordinates of the nuclei and find a new energy. This procedure is the basis for drawing potential curves or surfaces as in Figure 2.3. In Figure 2.3(a) the potential curve shows how the energy of H_2 varies with the internuclear distance; in Figure 2.3(b) the variation of the energy of H_3 with distance is shown. As there are two variables in this case, the two bond lengths are plotted on the x and y axes, and the energy is shown by contour lines.

C. The LCAO approximation

So far we have seen that the total wavefunction of a molecule can be separated into two parts, one part describing the nuclei and the other the electrons. The electronic part of the wavefunction is written as

$$\Psi = \phi_1\phi_2 \cdots \phi_n,$$

where each ϕ represents the wavefunction of a single electron and the wavefunction is 'antisymmetrized' in order to obey the Pauli principle. The question now is what form the orbitals ϕ take.

One of the commonest approaches to this problem is to express each orbital as a sum of atomic orbitals ($1s$, $2p$, etc.) based on the various atoms in the molecule. This is called the Linear Combination of Atomic Orbitals (LCAO), and may be written:

$$\phi_i = \sum_m c_{im}\chi_m,$$

(a)

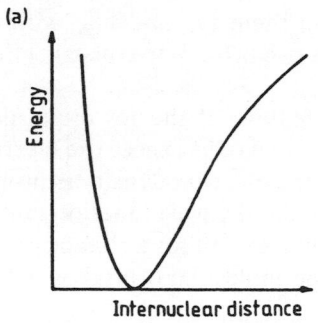

(b)

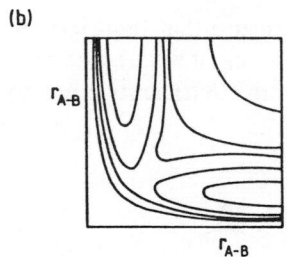

Figure 2.3 (a) The potential
energy curve for a diatomic
molecule. (b) The potential
energy surface for a triatomic
molecule

where χ_m is an atomic orbital and c_{im} is a mixing coefficient. Provided that enough atomic orbitals are considered, this can be a very accurate method of representing molecular orbitals. The values of mixing coefficients c_{im} are adjusted so that the energy of the molecule as calculated by the Schrödinger equation is as low as possible; a theorem known as the variation principle shows that the lower the calculated energy of the approximate wavefunction, the more accurately it represents the true wavefunction.

The accurate calculation of molecular orbitals as linear combination of atomic orbitals presents massive computational problems, and is feasible only for small molecules. However, the essential qualitative features of chemical bonding can be understood simply if the atomic orbitals from which the molecular orbitals are constructed are restricted to those which are occupied in the separate atoms.

D.   Bonding and antibonding molecular orbitals

Consider the hydrogen molecule, H_2. It contains two electrons; if the H–H bond were broken, we would have two hydrogen atoms, each with one electron in a $1s$ orbital. These $1s$ orbitals are identical to each other and have the form

$\chi = Ne^{-r}$. We shall label them $1s_A$ and $1s_B$, with A and B representing the two nuclei, and use these two orbitals to construct the molecular orbitals of the hydrogen molecule.

There is one further feature of the H_2 molecule that we can use: as the molecule is symmetrical, we would expect the electron density to be the same round each nucleus. As the electron density is proportional to the wavefunction squared, this means that the wavefunction must either be identical round each nucleus or be the same except for a change of sign.

We can now express our molecular orbitals ϕ in terms of the atomic orbitals $1s_A$ and $1s_B$:

$$\phi = c_A 1s_A + c_B 1s_B.$$

One way in which we can ensure that the electron density is the same round each nucleus is by taking $c_A = c_B$; if we are not concerned with normalizing our function, then we can put both coefficients equal to one, so that

$$\phi = 1s_A + 1s_B,$$

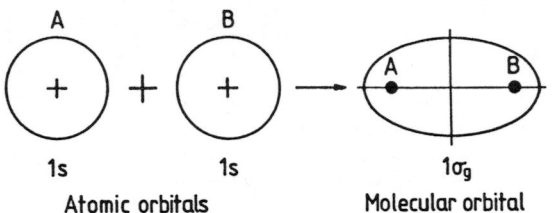

1s   1s $1\sigma_g$

Atomic orbitals   Molecular orbital

Figure 2.4 Formation of the $1\sigma_g$ molecular orbital

which is illustrated pictorially in Figure 2.4. This molecular orbital is called the $1\sigma_g$ orbital; it is a three-dimensional function which is positive everywhere. The notation σ indicates that it is cylindrically symmetrical about the A–B axis. The subscript 'g' stands for the German word *gerade*, even, meaning that the function is also symmetric with respect to the centre of symmetry. Figure 2.5 shows how the electron density, which is proportional to ϕ^2, compares with the electron density in two separate hydrogen atoms. The formation of the

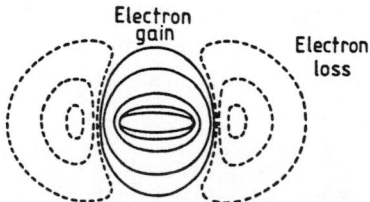

Figure 2.5 Electron density difference between H_2 and two individual H atoms with electrons in a $1\sigma_g$ molecular orbital

molecular orbital produces a build-up of negative charge between the positive nuclei; this is energetically favourable and so the $1\sigma_g$ orbital is called a *bonding* orbital.

There is one other way of combining the $1s_A$ and $1s_B$ orbitals so as to produce an electron density which is the same round each nucleus. If $c_A = -c_B$, then the sign of the wavefunction is positive round one nucleus and negative round the other, but the electron densities, which depend on ϕ^2, are the same. Thus

$$\phi = 1s_A - 1s_B.$$

This molecular orbital is called $1\sigma_u$ and is shown in Figure 2.6, where again σ indicates cylindrical symmetry, while the subscript 'u' (from the German word for odd, *ungerade*) refers to the function being antisymmetric—changing sign on reflection at the centre of symmetry.

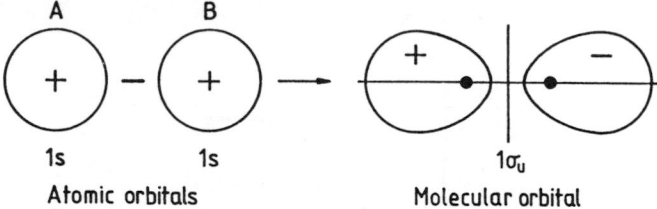

Figure 2.6 Formation of the $1\sigma_u$ molecular orbital

The electron density of the σ_u orbital is shown in Figure 2.7; the charge build-up is now away from the positively charged nuclei. This is energetically unfavourable, and the σ_u orbital is called an *antibonding* orbital. Note that it is still axially symmetrical and is therefore still a σ orbital.

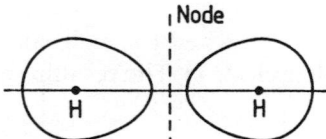

Figure 2.7 Electron distribution of the $1\sigma_u$ orbital of H_2

The formation of two molecular orbitals when two hydrogen atoms come together is illustrated in Figure 2.8; the separate atomic orbitals are shown on the left and right, and the molecular orbitals in the centre. In the H_2 molecule both electrons can occupy the bonding $1\sigma_g$ orbital, and the molecule is therefore more stable than two separate hydrogen atoms.

The amount by which the $1\sigma_g$ orbital is stabilized depends on how much the two hydrogen atomic orbitals overlap. If the distance between the nuclei is large, then there is little overlap and the molecule is not very stable. The stability of the electrons increases and the nuclei come closer together. However, the total energy of the molecule is the sum of the electronic energy

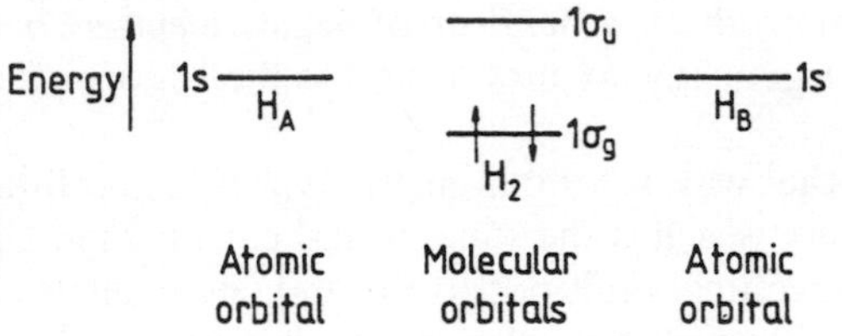

18

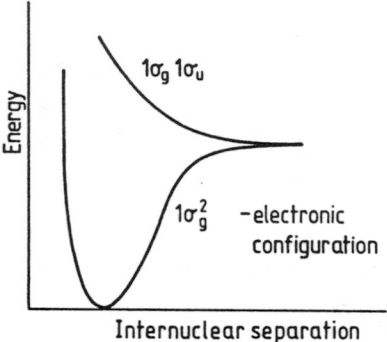

Figure 2.8 Molecular orbital diagram for H_2

and the nuclear-nuclear repulsion energy; this repulsion energy dominates at very short distances, and so the variation of the total energy with bond length for an H_2 molecule is shown in Figure 2.9.

Figure 2.9 Potential energy curves for the hydrogen molecule in its ground states

Promoting an electron to the excited $1\sigma_u$ molecular orbital will result in one bonding and one antibonding electron. The resulting molecular electronic state is unstable.

E. Molecular orbital diagrams

The notion of bonding and antibonding orbitals and building a picture of molecular orbital structure from constituent atoms can be extended very simply to other diatomics.

Diatomic lithium

The molecular orbital diagram for Li_2 is illustrated in Figure 2.10. Once again we have a preponderance of bonding over antibonding molecular orbitals with the consequence that Li_2 is a stable bound entity which is found in the vapour of heated lithium metal. In this case the $1s$ atomic orbitals are largely retained in the vicinity of the nuclei, being tightly bound so that it is the $2\sigma_g$

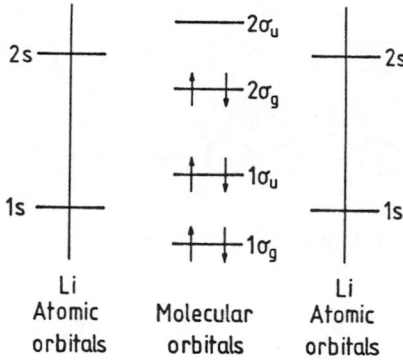

Figure 2.10 The molecular orbital diagram for Li_2

orbital which gives the single bond holding the molecule together. The extent to which a molecular orbital does contribute to bonding between two atoms is related to overlap of the separate atomic functions when they combine to form a molecular orbital. Antibonding molecular orbitals have no overlap—positive and negative portions cancelling. In the case of Li_2 the $2\sigma_g$ molecular orbital involves the overlapping of the spherical $2s$ atomic functions. This is not extensive so that the bond in Li_2 is not strong. An indication of this is the low value of the energy required to separate the atoms—the dissociation energy.

Mention of the dissociation energy reminds us of the fact that the use of orbitals and linear combinations to give molecular orbitals is an approximation and cannot give the complete picture. Illustrating this is the experimental fact that the dissociation of energy of diatomic lithium, which has an excess of two $2\sigma_g$ bonding orbitals, is less than that of Li_2^+ with only one. This apparently puzzling result reminds us that we are dealing with approximations and that, even though we may say that the structure of Li_2^+ may be represented as

$$\psi(Li_2^+) = 1\sigma_g^2\ 1\sigma_u^2\ 2\sigma_g,$$

this does not imply that $1\sigma_g$, $2\sigma_g$, etc., represent precisely the same mathematical functions in Li_2^+ as the same symbols do for Li_2. The symmetries must be the same in molecule and ion, but the energy may be significantly different.

F.   The use of 'p' orbitals in molecular orbital diagrams

When atoms with $2p$ electrons are involved in molecule formation, these atomic orbitals may also overlap to give bonding and antibonding molecular orbitals. It is conventional to describe the internuclear axis as the z axis of the molecule so that overlap of $2p_z$ functions on two atoms will yield a molecular orbital which is cylindrically symmetrical round the bond (Figure 2.11).

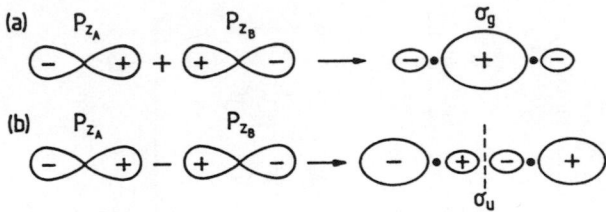

Figure 2.11 Molecular orbitals from atomic p_z functions

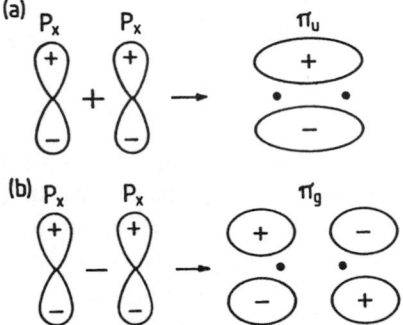

Figure 2.12 Linear combinations of
atomic $2p_x$ orbitals

A new aspect arises when the overlap is between the $2p_x$ and $2p_y$ atomic orbitals and is illustrated in Figure 2.12.

If the atoms A and B are identical the resulting molecular orbitals have a symmetry property with respect to the centre of symmetry of the molecule (i.e. the mid-point of the bond between the atoms). Points reflected through this point are identical locations (Figure 2.13). This reality must be reflected in the wavefunctions so that, for instance, since the electron density must be

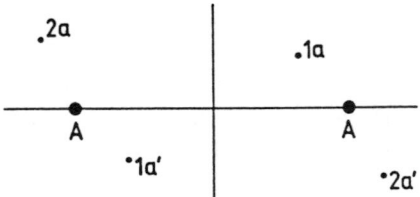

Figure 2.13 Reflection properties in molecules with a centre of symmetry: points a and a' are equivalent

identical for equivalent positions related by symmetry, ϕ^2, the square of the orbital must be the same at the two positions. Thus if ϕ^2 is the same at 1a and 1a', etc., then ϕ must be the same or only change sign at any two equivalent sites. This even or odd behaviour of the wavefunction is indicated in the notation by subscripts 'g' for even functions with no change of sign or 'u' for odd functions which change sign, just as for σ functions met above. If the two atoms are not identical and we have a heteronuclear molecule, then this symmetry is absent and the molecular orbitals are merely σ or π.

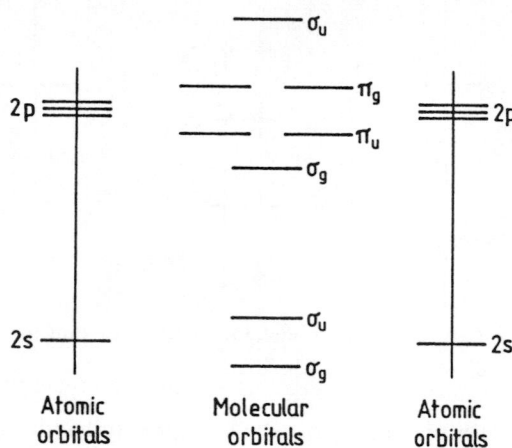

Figure 2.14 The order of molecular orbitals appropriate to the heavier homonuclear diatomic molecules

For atoms on the right-hand side of the periodic table, the 2s atomic orbitals are at very low energy with respect to the 2p orbitals and the order of energies of the molecular orbitals is as shown in Figure 2.14. This applies to the homonuclear molecules O_2 to Ne_2. For the left-hand part of the table, 2s and 2p are closer together in energy and the σ molecular orbitals interact so that the σ orbital largely derived from $2p_z$ atomic orbitals lies higher in energy than the bonding π level, as in Figure 2.15. This is the order found in the series Li_2 to N_2. We will now look at some examples.

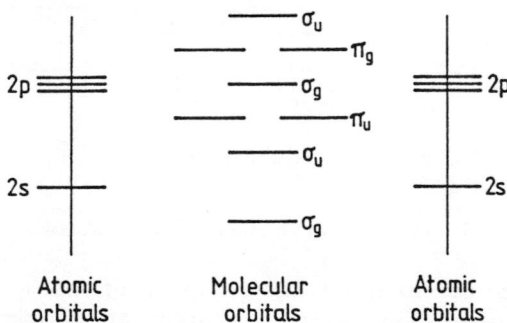

Figure 2.15 The order of molecular orbitals appropriate to the lighter homonuclear diatomic molecules

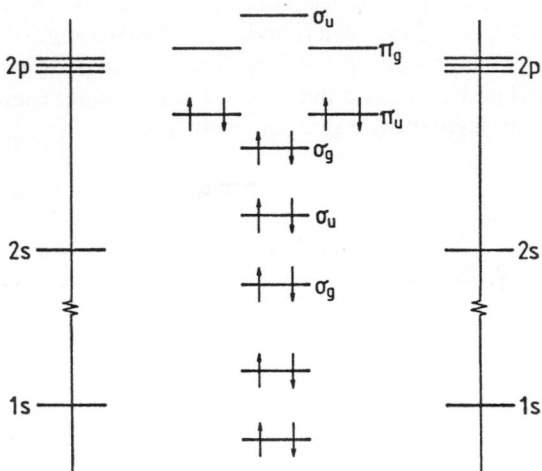

Figure 2.16 The molecular orbital diagram for N_2

Nitrogen

The molecular orbital diagram for the nitrogen molecule is shown in Figure 2.16. As may be seen, the π orbitals come in pairs with the same energy (or in the terminology of quantum mechanics they are degenerate). This arises because of the equivalence of overlapping two $2p_x$ orbitals on the constituent atoms or two $2p_y$. We could thus label the degenerate molecular orbitals as π_x and π_y. A theorem of quantum mechanics shows that if we have two degenerate solutions such as these, then any linear combination $c_1\pi_x + c_2\pi_y$ will likewise be a solution. This includes the possible combinations $(\pi_x + i\pi_y)$ and $(\pi_x - i\pi_y)$ with the coefficient i being $\sqrt{-1}$. This seemingly odd choice does have some importance because the simplest form of the atom $2p$ functions has them incorporating complex numbers. The two combinations are frequently labelled π^+ and π^-, so that π orbitals have alternative labellings which do not influence the energy (Figure 2.17). Although the real forms π_x and π_y

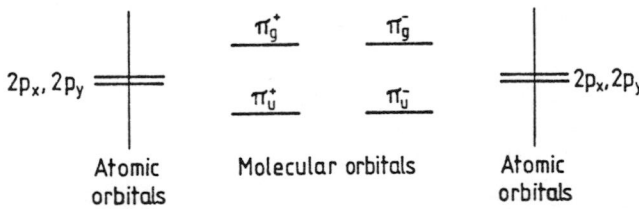

Figure 2.17 The alternative labelling of π molecular orbitals

may be more easily visualized, the complex alternatives π^+ and π^- are more useful when considering angular momentum (compare $2p_x\,2p_y\,2p_z$ and $2p_+$, $2p_-$, $2p_0$ in the atomic case). Although not strictly correct, it is often helpful to think of π electrons having one unit of angular momentum about the internuclear axis with π^+ moving in one direction and π^- in the other. If we then fill a π shell with four electrons using the *aufbau* principle, then there will be no resultant angular momentum as two are π^+ and two are π^-, with two having α spin and two β.

Carbon Monoxide

So far the diatomic molecules we have met have had identical nuclei and are said to be *homonuclear*. For a heteronuclear molecule such as CO with the same number of electrons as N_2, the only major difference is that the molecular orbital coefficients are not determined any longer by symmetry.

The tightly bound molecular orbitals will be as before:

$$\phi = c_1 1s_C + c_2 1s_O,$$

but now there is no simple relationship between c_1 and c_2. Since oxygen has a higher nuclear charge than carbon, we may expect that c_2 is much larger than c_1 and that the molecular orbital is in consequence rather like an atomic $1s$ oxygen function. It will still have symmetry round the axis and can thus be called a σ orbital but the g and u inversion symmetry will no longer be present. The useful simplified molecular orbital diagram is shown in Figure 2.18.

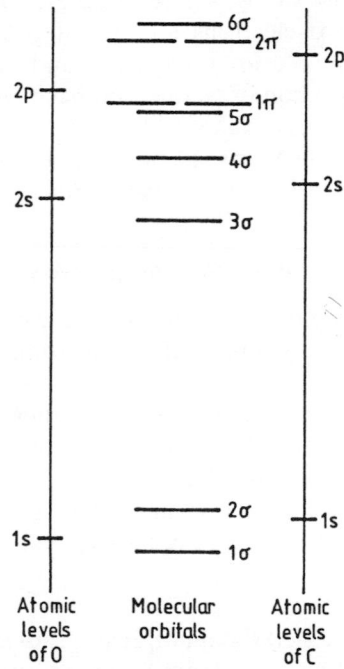

Figure 2.18 Molecular orbitals for CO

Oxygen

Oxygen is rather like nitrogen except that it has two extra electrons, so we can go directly to the molecular orbital diagram of Figure 2.19. The manner in

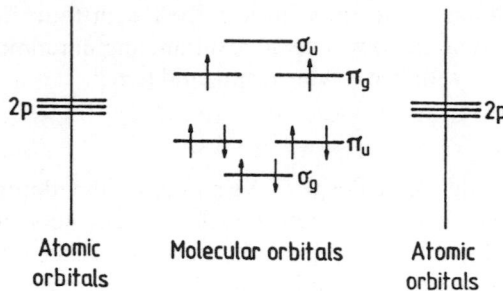

Atomic orbitals   Molecular orbitals   Atomic orbitals

Figure 2.19 The higher occupied molecular orbitals in O_2

which the available molecular orbitals have been filled in this diagram indicates once again just how much we can carry over from our study of atoms to the molecular case.

Firstly we have always put at most two electrons in each molecular orbital: one with α spin and one with β (indicated by arrows pointing up or down). This is the Pauli principle.

Secondly, as in the case of O_2, where there are alternative ways of filling orbitals as in putting two electrons into the degenerate π_g pair, we have followed the atomic example for p electrons and put one electron in each orbital to maximize separation of charge and have given them parallel spins. This is an application of Hund's rules.

Thus molecular orbital diagrams can be used just like atomic energy levels. We carry over the *aufbau* principle, the Pauli principle and Hund's rules.

G.   Designation of diatomic electronic energy levels

In the case of atoms the angular momenta of the constituent electrons couple to provide a resultant which serves to label the electronic state. There is a similar situation in the case of molecules.

If we have a closed-shell pairing of all electrons then there is no resultant spin or angular momentum. The electronic state will be a singlet, and if the molecule is a diatomic with no resultant orbital angular momentum, we write the 'name' of this energy level (or its term symbol) as $^1\Sigma$. Paralleling the atomic case, the general form of the term symbol is written as

$$^{2\Sigma+1}\Lambda, \text{ e.g. } ^1\Sigma, {}^3\Pi, {}^2\Delta.$$

The top left superscript, called the multiplicity, is twice the value of the total spin (called Σ corresponding to total spin S in atoms) plus one. Again the origin of this seemingly unnecessarily complex system is historic but it persists because it does bear some relation to the appearance of electronic spectra. The resultant Λ is the angular momentum about the internuclear axis. Electrons of σ type contribute nothing, π electrons one unit, δ electrons two units and, as in atoms, the constituents are added vectorially.

Two further additions may be made to the diatomic term symbol. For homonuclear molecules a subscript 'g' or 'u' is added to indicate whether the total wavefunction is symmetric or antisymmetric with respect to inversion at the coordinate origin or equivalenty to interchange of the identical nuclei. This may be ascertained by inspection from the molecular orbital structure since states are of type u if the number of u electrons is odd and g if there are an even number of u electrons. Thus the ground state of oxygen which we have just encountered is

$$\Psi(O_2 \text{ ground state}) = 1\sigma_g^2\, 1\sigma_u^2\, 2\sigma_g^2\, 2\sigma_u^2\, 3\sigma_g^2\, 1\pi_u^4\, 1\pi_g^2$$

and has a term symbol $^3\Sigma_g$.

The other addition to the term symbol refers to the symmetry of the wavefunction with respect to a plane through the nuclei, and applies both to homonuclear and heteronuclear diatomic molecules. States which are symmetric are labelled with a superscript plus and antisymmetric states with a minus. Again this symmetry property is revealed by inspection of the molecular orbital structure, but we must be careful to use π^+ and π^- rather than real forms and to recall that the molecular orbital 'structure' is really shorthand notation for a determinantal function. If we change π^+ to π^-, and vice versa, we will thus alter columns of a determinant, and single changes cause a sign change (antisymmetry) whereas an even number of changes leaves an equivalent determinant (symmetric). Thus in full the oxygen ground state term symbol is

$$^3\Sigma_g^-$$

because the determinantal function is

$$\Psi = 1\sigma_g^2\, 1\sigma_u^2\, 2\sigma_g^2\, 2\sigma_u^2\, 3\sigma_g^2\, 1\pi_u^{+\alpha}\, 1\pi_u^{+\beta}\, 1\pi_u^{-\alpha}\, 1\pi_u^{-\beta}\, 1\pi_g^{-\alpha}.$$

On the other hand, the ground states of H_2 and N_2 are both states of symmetry

$$^1\Sigma_g^+.$$

Knowledge of the details of these term symbols is not very important but the notation is constantly encountered and introduced here in the hope that unfamiliar notation will not be misconstrued as genuine complexity.

H.   Summary

The electronic structure of diatomic molecules is very similar to that of atoms except that the symmetry is cylindrical rather than spherical. The molecular orbitals are most easily visualized as combinations of the atomic orbitals from the bonded atoms. The actual bond is a concentration of negative electronic charge which holds together the positive nuclei. The rotations of the molecule as a whole or the vibrations of the bond give rise to extra types of energy level in addition to the electronic energy found in atoms. These energy levels will be the subject of Chapter 4, after a discussion of the electronic structure of polyatomic molecules.

I.   Problems

1.   Comment on the following: H_2^+ has a longer bond length than H_2, but O_2^+ has a shorter bond length than O_2.
2.   Why does the molecule He_2 not exist? Would you expect He_2^+ to be stable to dissociation?
3.   What are the molecular orbital configurations of Li_2^+, NO and NF in their ground electronic states?
4.   Draw a molecular orbital diagram for F_2, labelling the orbitals correctly. How many more electrons are there in bonding orbitals than anti-bonding orbitals?
     What would the spectroscopic term symbols be for the states of F_2 with one electron excited from the bonding π molecular orbital to the corresponding antibonding π?
5.   The diatomic species OH was discovered in interstellar space in 1964. Describe its molecular orbital configuration. What can simple m.o. theory tell us about the nature and location of the unpaired electron?

Chapter 3

Electronic Structure of Polyatomic Molecules

A. Symmetry properties of orbitals

In the case of diatomic molecules which were considered in the last chapter, we saw that for homonuclear cases such as H_2 or N_2, the molecular orbitals have symmetry properties. We write the molecular orbital as a linear combination of atomic orbitals,

$$\phi = c_1\chi_A + c_2\chi_B,$$

where χ_A and χ_B are identical atomic orbitals centred on the two nuclei A and B. If the nuclei are identical then, of course, their interchange would make no difference to the electron density, i.e. ϕ^2 at any point in space must be invariant to the exchange of nuclei. This being so, ϕ must either remain exactly the same or just change sign at any point, and we labelled these two possibilities 'g' or 'u'. Similarly, all diatomic molecules have a plane of symmetry (Figure 3.1). Again, charge density above and below the plane must be identical so we have orbitals such as σ or π, the former being indentical mirror images above and below the plane while π orbitals change sign on opposite sides.

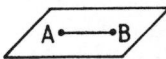 Figure 3.1 A plane of symmetry in a heteronuclear diatomic molecule

These simple ideas may be generalized and form the basis of group theory. Group theory is particularly useful in polyatomic molecules, but in this introduction we can only hint at its power by considering some simple examples, namely H_2O and NH_3.

B. Water, H_2O

In the notation of group theory H_2O is said to have C_{2v} symmetry, denoting the presence of two symmetry planes and a two-fold rotation axis (C_2) through the

27

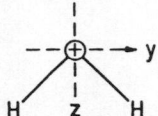

Figure 3.2 Axes in the structure of the water molecule

oxygen atom in the z direction (Figure 3.2). With these symmetry elements the electron density must be invariant to making a mirror image of the molecule so that all the molecular orbitals must be symmetric or antisymmetric (stay the same or only change sign) with respect to each sort of symmetry operation. The notation employed is as follows:

Molecular orbitals of types a_1 and a_2 are unchanged by the C_2 rotation.

Molecular orbitals of types b_1 and b_2 change sign on C_2 rotation.

Molecular orbitals of types a_1 and b_2 are unchanged by reflection of the molecule in the yz plane.

However,

Molecular orbitals of types a_2 and b_1 do change sign on this reflection.

We may use these symmetry labels to 'name' the molecular orbitals.

If the molecular orbitals (m.o.) are made up by a linear combination of atomic orbitals (LCAO) then we can see that the following atomic orbitals (a.o.) or combinations thereof have the symmetry properties just described:

m.o.	a.o.	
a_1 symmetry	$1s_O$	Oxygen atomic orbitals
	$2s_O$	
	$2p_{z_O}$	
a_1 symmetry	$(1s_{H_1} + 1s_{H_2})$	Hydrogen atomic orbitals
b_2 symmetry	$2p_{y_O}$	
	$(1s_{H_1} - 1s_{H_2})$	
b_1 symmetry	$2p_{x_O}$	

The atomic orbital with the lowest energy is the oxygen $1s$ followed by oxygen $2s$. Hence the molecular orbitals have largely the following composition:

$1a_1$ mostly $1s_O$

$2a_1$ mostly $2s_O$

$3a_1$ mostly $1s_H + 1s_H$

$1b_2$ mostly $2p_{y_O}$

$1b_1$ mostly $2p_x$, a non-bonding m.o.

The molecular orbital structure of water can thus be written as

$$1a_1^2 \ 2a_1^2 \ 1b_2^2 \ 3a_1^2 \ 1b_1^2.$$

This procedure follows exactly in parallel with electronic configurations already encountered for atoms or diatomic molecules. As in those cases, the symmetry arguments do not give the actual order of energies, which can only come from detailed calculation or a rationalization of the electronic spectrum where electrons change orbitals.

C. Ammonia, NH_3

Figure 3.3 shows the shape of the ammonia molecule and illustrates the fact that the three hydrogen atoms form an equilateral triangle with three symmetry planes (marked σ) and a threefold rotational symmetry axis (C_3).

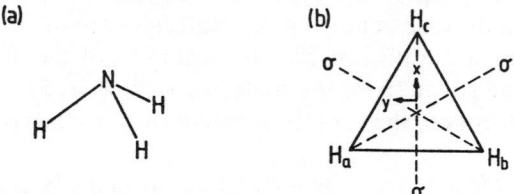

Figure 3.3 (a) The ammonia molecule. (b) Symmetry of the hydrogen atoms

Once again the molecular orbitals must change sign or stay the same when any symmetry operation is performed, and the behaviour with respect to each operation defines a symmetry and provides a name:

m.o.	a.o.

a_i symmetry
    no change of sign
    for σ or C_3
	$1s_N$
	$2s_N$
	$2p_{z_N}$
	$1s_A + 1s_B + 1s_C$

e symmetry
    change of sign for C_3.
    These orbitals cannot be
    obtained by inspection
    and come in pairs like
    π molecular orbitals in
    diatomics
	$\begin{cases} 2p_x^N \\ 2(1s_{N_C}) - 1s_{H_A} - 1s_{H_B} \end{cases}$
	
	$\begin{cases} 2p_y^N \\ 1s_{H_A} - 1s_{H_B} \end{cases}$

The molecular orbital configuration of ammonia in its electronic ground state is then

$$1a_1^2 \ 2a_1^2 \ 1e^4 \ 3a_1^2,$$

30

with

    $1a_1$—mostly $1s_N$
    $2a_1$—mostly $2s_N$
    $1e$—a bonding mixture of several atomic orbitals
    $3a_1$—mostly $2p_{z_N}$, the lone pair.

Polyatomic molecular orbital configurations are useful in explaining the spectra of the molecules, exactly paralleling the electronic configuration of atoms and diatomic molecules. The only complication is the wide variety of symmetry types possible. Molecular orbital diagrams, too, are of tremendous help.

D.   Walsh diagrams

The molecular orbital diagrams met earlier for diatomic molecules provide help in understanding the properties, both spectroscopic and chemical, of diatomics. Similarly useful in the case of polyatomic molecules are the Walsh diagrams.

    Walsh diagrams illustrate just how the energies of the various molecular orbitals vary as the geometry of the molecule is altered. They may be used to predict the most favoured shape of the molecule in both the ground and excited electronic states.

    All the essential features may be understood from the Walsh diagram for H_3 shown in Figure 3.4. The three molecular orbitals labelled $1a_1$, $1b_2$ and $2a_1$ become σ_g and σ_u symmetry in the linear case.

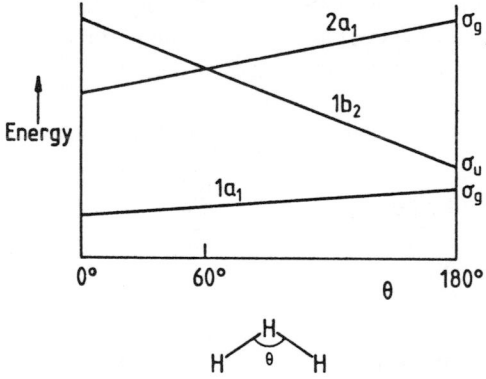

Figure 3.4 Schematic Walsh diagram for H_3

    If we now feed in electrons, H_3^+ will have a structure $1a_1^2$. The diagram suggests that this would be most stable with the angle $\theta = 0°$. Clearly nuclear repulsion will prevent this and the best compromise is $\theta = 60°$—the three hydrogen atoms at the corners of an equilateral triangle.

    In H_3 with its third electron the slope of the $1b_2$ orbital energy as a function of θ indicates that the energetically most stable situation will be linear. Experimentally these two structural conclusions are verified.

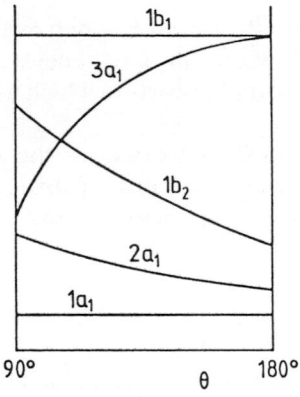

Figure 3.5 Schematic Walsh
diagram for AH_2 molecules

Diagrams similar to this can be constructed for many small polyatomic types. For instance Figure 3.5 is the Walsh diagram for molecules of general formula AH_2. Conclusions in this case are that six-electron species (BeH_2, BH_2^+) should be linear, but those with seven, eight, nine or ten electrons should be bent. Thus BH_2 with one $3a_1$ electron has a bond angle of 131°; CH_2 with two $3a_1$ electrons is more bent with an angle of 103°; H_2O has a bond angle of 104.5° in harmony with the diagram.

Whether excitation of an electron will favour increasing or decreasing the bond angle can be determined from the diagrams. Clearly it will depend on the variation of the energy of the orbital into which the electron is excited. Thus for BeH_2

$1a_1^2\ 2a_1^2\ 1b_2^2$ is linear,
$1a_1^2\ 2a_1^2\ 1b_2\ 3a_1$ is bent and
$1a_1^2\ 2a_1^2\ 1b_2\ 1b_1$ is again linear.

Analogous diagrams can be drawn for a wide variety of different structures. They cannot be rigorously justified but are immensely useful.

E.   The Jahn–Teller effect

Consideration of the simple Walsh diagram for H_3^+ enables us to understand a fascinating effect which results from the Jahn–Teller theorem. If we take H_3^+ at 60° and excite an electron it can go into either the $2a_1$ or $1b_2$ orbitals, which have the same energy at the regular angle of 60°. However, a distortion from 60° must produce an improvement in energy whichever way we distort, as a crossing of orbital energies is only possible if there is a negative slope on either side. Thus this excited state distorts and is not a regular equilateral triangle.

The generalization of this effect can be stated as the Jahn–Teller theorem: 'Any non-linear molecular system in a degenerate electronic state will be unstable and undergo some sort of distortion which will lower its symmetry and split the degenerate state.'

This effect is probably most important in the case of transition metal complexes, which we will meet at the end of this chapter. For example, the complex ion $[CoF_6]^{3-}$ is not a regular octahedron.

F. Hybridization

So far we have become accustomed to forming a molecular orbital, ϕ, by the combination of two atomic orbitals, χ:

$$\phi = c_1\chi_A + c_2\chi_B.$$

We now consider what is likely to happen if perhaps an orbital on atom B is close in energy to two orbitals of atom A and moreover is of the correct symmetry to overlap. An example is the case of the $1s$ orbital of hydrogen and the $2s$ and three $2p$ orbitals of carbon. Figure 3.6 illustrates the five molecular orbitals which may arise from combining the five atomic functions.

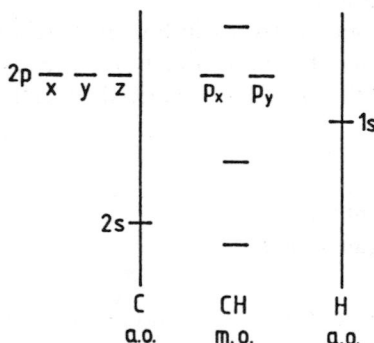

Figure 3.6 Molecular orbital diagram for CH

An alternative way of viewing this is to mix the $2s$ and $2p_z$ on the carbon first, as in Figure 3.7. Here sp_+ and sp_- are called 'hybrids' and look like the functions shown in Figure 3.8. One of these two hybrids can overlap with an approaching hydrogen $1s$ atomic orbital as in Figure 3.9. This yields a set of molecular orbitals identical to those illustrated in Figure 3.5, which emphasizes the point that we can form hybrids or not as we please and the final result will not be affected by our choice.

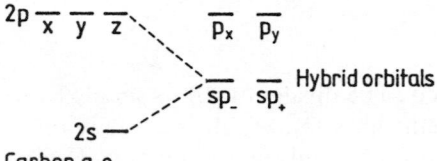

Figure 3.7 Formation of *sp* hybrid orbitals

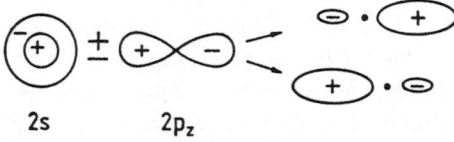

Figure 3.8 Formation of *sp* hybrid orbitals

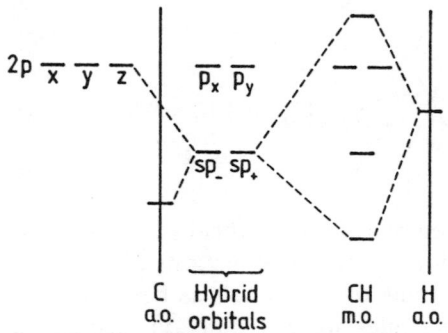

Figure 3.9 Formation of molecular orbitals following hybridization

G. The shape of water

If we now look again at water we can understand the shape of the molecule better if we think in terms of hybridization. The atomic structure of oxygen is

$$1s^2\ 2s^2\ 2p_z^2\ 2p_x\ 2p_y.$$

This might lead to the expectation that if bonds are formed between the $2p_x$ and $2p_y$ orbitals on oxygen overlapping with $1s$ orbitals on each of the hydrogens, then the bond angle should be 90°. In fact the measured angle is about 104° and the explanation is that some oxygen $2s$ is also involved and may be described as 'hybridization'.

H. Methane—valence states

Organic chemistry is an area where the ideas of hybridization are most often used. Carbon has the atomic structure

$$1s^2\ 2s^2\ 2p_x\ 2p_y$$

and might be expected to be divalent with its simple hydride being CH_2. As is well known, it is methane, CH_4, which is actually the common compound, known to be tetrahedral since only one isomer of CH_3Cl occurs.

To obtain CH_4 the carbon has to be notionally promoted to a valence state with the configuration

$$1s^2\ 2s\ 2p_x\ 2p_y\ 2p_z,$$

which requires 436 kJ mol^{-1}. This is compensated for by the added stability of the four covalent bonds produced in methane, each with a bond energy of approximately 450 kJ mol^{-1}. The valence state does not occur for the isolated atom but is a hypothetical condition through which it must pass.

Combining a $2s$ atomic orbital with three $2p$ functions gives four 'sp^3' hybrids, tetrahedrally distributed as in Figure 3.10.

Figure 3.10 sp^3 hybrid orbitals

It is important to emphasize that hybridization is not an effect but rather an alternative way of looking at a wavefunction. We have already seen how in the atomic case we can think of 'p' orbitals as being p_x, p_y and p_z or p_+, p_0 and p_-. In diatomic molecules we can think in terms of π_x and π_y molecular orbitals or π^+ and π^-. Now for carbon compounds we may think of carbon having an electronic structure of s and p^3 or as having four sp^3 hybrids. Perhaps the most clear-cut demonstration that these are only alternative ways of looking at the same thing comes by recalling that our wavefunctions are really determinantal, i.e.

$$\begin{vmatrix} s(1) & p_z(1) & p_y(1) & p_x(1) \\ s(2) & p_z(2) & p_y(2) & p_x(2) \\ s(3) & p_z(3) & p_y(3) & p_x(3) \\ s(4) & p_z(4) & p_y(4) & p_x(4) \end{vmatrix} \equiv \begin{vmatrix} t_1 & & & \\ & t_2 & & \\ & & t_3 & \\ & & & t_4 \end{vmatrix}$$

where

$$t_1 = s + p_x + p_y + p_z$$
$$t_2 = s - p_x - p_y + p_z$$
$$t_3 = s + p_x - p_y - p_z$$
$$t_4 = s - p_x + p_y - p_z$$

This can be verified by multiplying out each of the alternative forms of the determinant when identical results are obtained.

It is easier to think in terms of hybrids, and the reality is not the wavefunction but the electron density, which is again identical whichever description we employ.

I. Other hybridizations

Classical chemistry has shown that carbon can also form double and triple bonds. In single bonds such as those found in methane we visualize hybrids of the sp^3 type. For double bonds one s and two p functions are combined to give planar sp^2 hybrids with the remaining p orbital in the perpendicular direction. Triple bonds arise when a single sp hybrid is considered, leaving two p orbitals in the mutually perpendicular planes. These structures are typified by ethene (ethylene) and ethyne (acetylene).

Ethene

In C_2H_4, each C atom forms a planar sp^2 hybrid as in Figure 3.11. The repulsion between the electron densities in the various bonds results in the actual HĈH angle being 117° rather than the regular trigonal value of 120°.

(a)

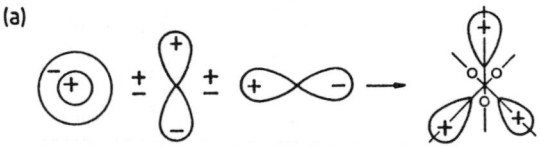

(b)

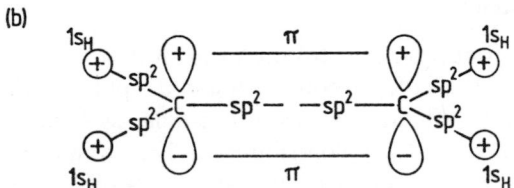

Figure 3.11 (a) Formation of sp^2 hybrids. (b) Bonding in ethene

Ethyne

In C_2H_2, each carbon forms the sp_+ and sp_- linear pair of hybrids, with the resulting structure as in Figure 3.12.

It is also possible to use d atomic orbitals, especially in the case of organic complexes. Table 3.1 summarizes the possible hybridizations and the resulting geometric arrangements.

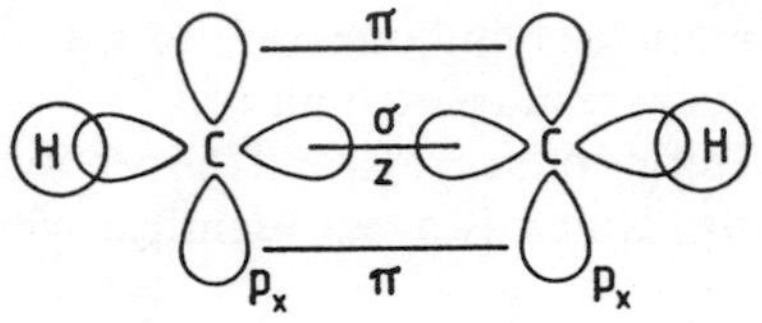

Figure 3.12 Bonding in ethyne

Table 3.1 Summary of hybridizations

Coordination number	Atomic orbital	Resulting hybrid geometry
2	sp	Linear
	dp	Linear
	sd	Bent
3	sp^2	Trigonal plane
	dp^2	Trigonal plane
	ds^2	Trigonal plane
	d^2p	Trigonal pyramid
4	sp^3	Tetrahedral
	d^3s	Tetrahedral
	dsp^2	Tetragonal plane
5	dsp^3	Bipyramidal
	d^3sp	Bipyramidal
	d^4s	Tetragonal pyramid
6	d^2sp^3	Octahedral
	d^4sp	Trigonal prism

J. Pi electron systems

In ethene the 'p' orbitals on the carbon atoms not involved in hybridization form the π bond between the carbons with a nodal plane in the plane of the σ C—C bond. A similar situation arises in benzene, as Figure 3.13 illustrates. Each of the six carbon atoms contributes a $2p_z$ atomic orbital which can combine to yield six π molecular orbitals into which six electrons must be fed. Rather in the manner of building a molecular orbital diagram for diatomics, the π molecular orbital diagram can be constructed qualitatively on pure symmetry grounds. The possible molecular combinations are shown in Figure 3.14, where each m.o. has to be symmetric or antisymmetric to one of the symmetry planes. The resulting energy diagram shows the orbitals increasing in energy as the number of nodes in the wavefunction increases, as is generally the case—even in simple systems such as the particle in a box.

In Figure 3.15 the six electrons are placed in the lower molecular orbitals; the reason for the stability of benzene is clear. For any similar conjugated ring

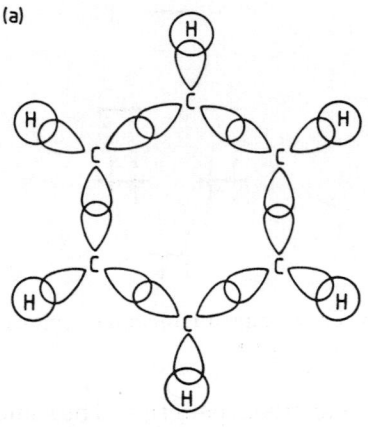

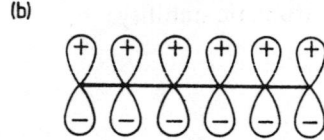

Figure 3.13 Molecular orbitals in benzene. (a) xy plane, (b) z direction

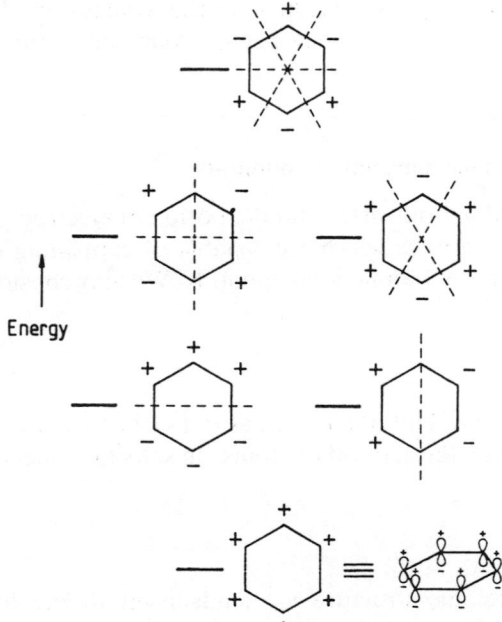

Figure 3.14 Schematic representation of the π-molecular orbitals of benzene

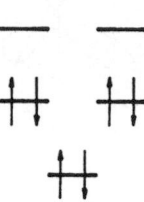

Figure 3.15 The occupied molecular orbitals of benzene

system the energy level and orbital pattern will be similar, with the lowest level unique and higher levels doubly degenerate, which is the origin of the $(4n + 2)\,\pi$ electron rule for aromatic stability.

K. Electron repulsion

A nice simple rationalization of the shapes of many polyatomic compounds is given by the ideas of electron repulsion formulated by Gillespie and Nyholm. Quite simply, electron pairs are thought to repel each other and maximize separation. If the pair is involved in a bond then the positive nucleus will reduce the negative charge density leading to the conclusion that lone pairs of electrons repel more than bonded pairs. Examples of this are illustrated in Figure 3.16.

L. Valence rules for inorganic compounds

Using the ideas of hybridization and the notion of electron pair repulsions we can formulate some rules which are capable of explaining many of the facts about the structures of inorganic compounds. We may consider these group by group.

Group I

These form weak covalent bonds such as in Li_2, but are found as metals, the s orbitals overlapping those of other atoms. In salts the s electron is transferred giving M^+.

Group II

The atomic electronic structure ns^2 lends itself to the formation of M^{2+} ions but a valence state sp gives rise to MX_2 linear compounds. In the case of barium, $6s$ to $5d$ promotion becomes more likely and the sd hybrids such as those in BaF_2 produce bent geometries.

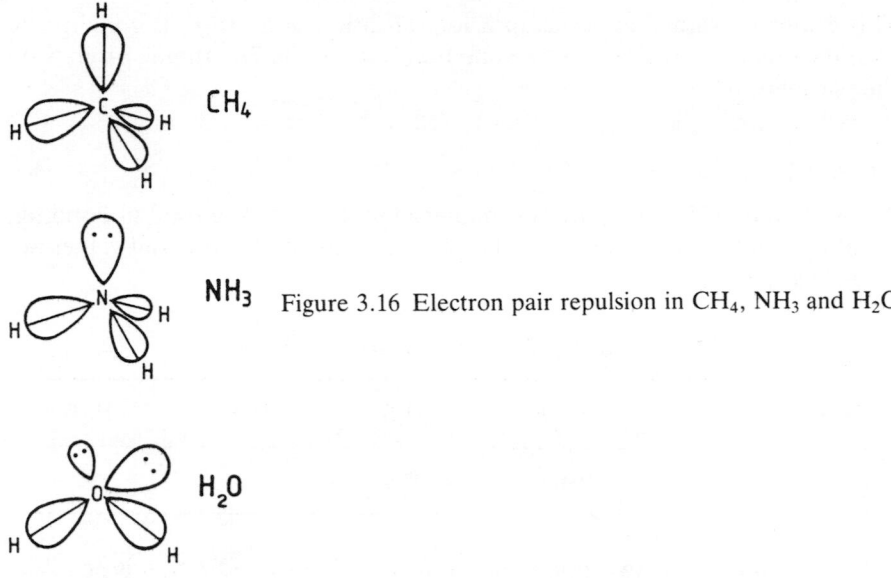

CH$_4$

NH$_3$ Figure 3.16 Electron pair repulsion in CH$_4$, NH$_3$ and H$_2$O

H$_2$O

Group III

The s^2p configuration does give some monovalent compounds such as BH, but the valence state sp^2 gives trigonal hybrids, e.g. BF$_3$. However, adding F$^-$ to BF$_3$ leads to BF$_4^-$ which is tetrahedral and in this form has eight electrons surrounding the nucleus. This tendency is so strong that BH$_3$ does not exist but B$_2$H$_6$ does.

Group IV

As we have seen, the carbon structure s^2p^2 gives rise to a range of hybridization possibilities.

Group V

The s^2p^3 configuration would give trivalent compounds with bond angles of 90° were hybridization not to take place. PH$_3$ has this approximate structure. Pentavalent compounds are also formed as in PCl$_5$ (Figure 3.17).

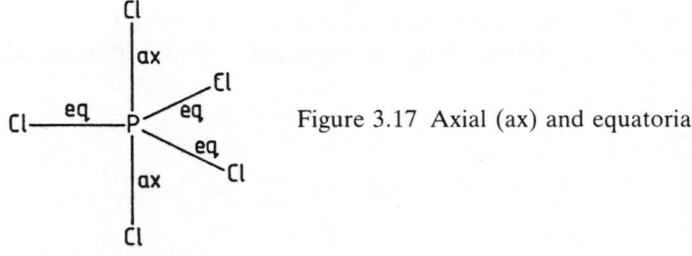

Figure 3.17 Axial (ax) and equatorial (eq) bonds in PCl$_5$

40

This can be explained either as sp^3d hybridization or an sp_xp_y trigonal plane with the other chlorine atoms forming bonds using the $3p_z$ atomic orbitals on the phosphorus.

Group VI

The s^2p^4 configuration, if the two unpaired p electrons are used in bonding, should give 90° bond angles for XH_2. The experimentally observed values are given in Table 3.2.

Table 3.2 Bond angles in group VI dihydrides

Molecule	H_2O	H_2S	H_2Se	H_2Te
Angle	104.5°	93°	91°	89.5°

Water could be the exception due to the fact that hybridization is possible, but another explanation would be the repulsion of the two hydrogen atoms which are closer together than those of the higher homologues.

Group VII

The halogens do form the obvious monovalent compounds such as F_2, HF and ClF. Since the energy separation between atomic s and p electrons is large, very little hybridization occurs.

An interesting variation is found in compounds such as ClF_3 and BrF_3 (Figure 3.18). This structure can be rationalized if we think of ionizing a $3p$ electron giving a 90° Cl^+F_2 geometry with the electron going to a third

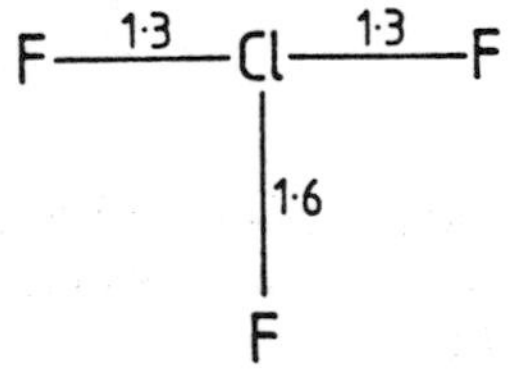

$$F \overset{1\cdot3}{\text{——}} Cl \overset{1\cdot3}{\text{——}} F$$
$$\Big|{\scriptstyle 1\cdot6}$$
$$F$$

Figure 3.18 The structure of ClF_3

fluorine atom giving F^-. The actual structure would be based on that shown in Figure 3.19 but by symmetry both 'horizontal' bonds would be equal in length.

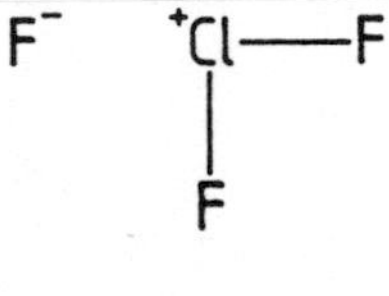

$$F^- \qquad {}^+Cl \text{——} F$$
$$\Big|$$
$$F$$

Figure 3.19 A resonance structure for ClF_3

Group VIII

Generally speaking the rare gases do not form compounds easily, but compounds such as XeF_2 can be formed from the heavier members of the group. XeF_2 is linear with explanations similar to those for ClF_3 being invoked to interpret the geometry.

M. Transition metal complexes

These invariably involve the d electrons on the metal atom, a typical example being $[Co(NH_3)_6]^{3+}$ which is an octahedral complex (Figure 3.20). If the ligands in this complex are considered to be point centres of electron repulsion then the d orbitals on the central atom will fall into two sets. The $d_{x^2-y^2}$ and d_{z^2} will point directly at ligands while d_{xy}, d_{yz} and d_{zx} point in between. The electron repulsion differences will thus split the degeneracy of the atomic d levels as in Figure 3.21.

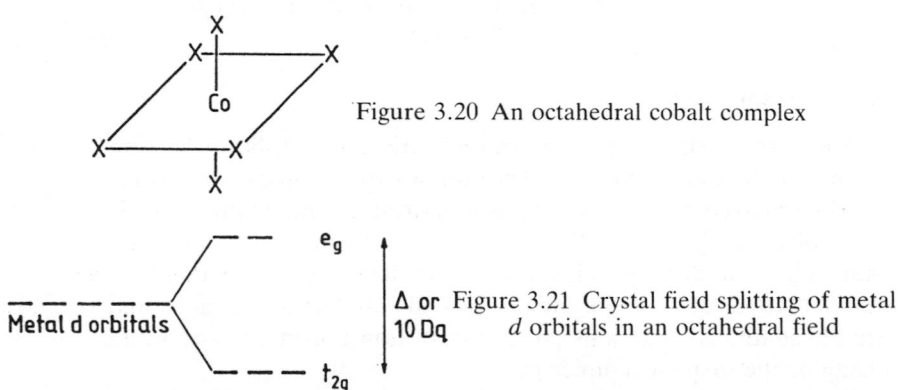

Figure 3.20 An octahedral cobalt complex

Figure 3.21 Crystal field splitting of metal d orbitals in an octahedral field

If we now apply the *aufbau* ideas to this diagram then when we reach the d^4 configuration we will have a choice of putting the fourth electron into one of the lower (t_{2g}) levels with the opposite spin to the one already there, or keeping parallel spins and putting it into the higher energy e_g level. The former alternative has two unpaired electrons while the latter has four, and these are referred to as high and low spin respectively. Which alternative is followed depends on the separation Δ (or 10 Dq) which in turn is a function of the nature of the ligands and the crystal field which they create. High spin and low spin complexes will clearly have very different magnetic and optical properties.

The explanation in the previous paragraph is often called 'crystal field theory'. A more satisfactory version called 'ligand field theory' is based on the use of a molecular orbital diagram. We have to consider that the molecular orbitals formed by overlap of the orbitals on two of the ligands with two of the five d atomic orbitals giving bonding and antibonding combinations as in

42

Figure 3.22. Into the molecular orbitals formed in this way we would feed the twelve electrons (two from each ligand) and again obtain an orbital structure with the possibilities of high and low spin complexes.

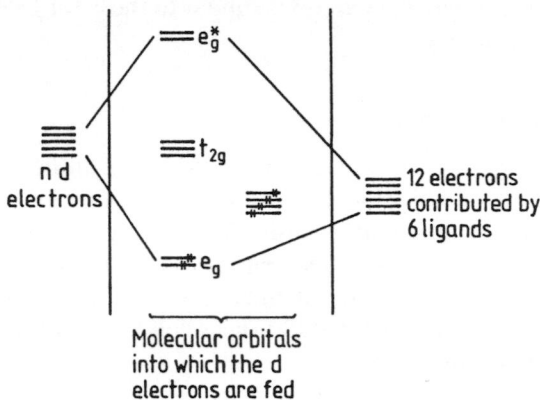

Figure 3.22 Ligand field structure of an octahedral complex

N. Summary

The electronic structure of polyatomic molecules follows on directly from atoms and diatomic molecules. The atomic orbitals on individual atoms may be considered to combine to give molecular orbitals with symmetries which reflect the molecular symmetry. Equivalent positions in a molecule must have the same electron density. This fact, formalized as group theory, facilitates understanding of how the atomic orbitals contribute to molecular orbitals. Simple qualitative diagrams can also show how the energies of orbitals vary on changing the shape of a molecule.

For large molecules simplified pictures such as hybridization schemes or maximizing electron–electron repulsion can give a very good idea of the shape of the ground electronic states of molecules.

O. Problems

1. (a) By calculating the number of pairs of electrons round one central nucleus, predict the shapes of the following:
    CH_3^+, CH_3^-, SF_6, ClF_3, BF_3NH_3.
    (b) Which of the following molecules would you expect to be linear
    NO_2, NO_2^+, NO_2^-, H_2O^+, H_2O^-?
2. Why do H_3O^+ and NH_4^+ exist, but not CH_5^+? Why does SF_6 exist, but not OF_6?
3. Why does N_2H_2 exist in two isomeric forms, while C_2H_2 has only one form?
4. The π-orbitals in cyclobutadiene are formed by overlap of four atomic

p-orbitals, ϕ_1, ϕ_2, ϕ_3 and ϕ_4, where the atoms are numbered cyclically. The resulting molecular orbitals are: $(\phi_1 + \phi_2 + \phi_3 + \phi_4)$, $(\phi_1 + \phi_2 - \phi_3 - \phi_4)$, $(\phi_1 - \phi_2 - \phi_3 + \phi_4)$ and $(\phi_1 - \phi_2 + \phi_3 - \phi_4)$. Sketch these molecular orbitals, and place them in order of energy. Would cyclobutadiene be expected to have a singlet or triplet ground state?

5. Predict the structure of compounds formed between xenon and fluorine.

Chapter 4
Energy Levels

In the case of atoms the energy level diagram can be presented as a set of horizontal lines on an energy diagram, each line representing an electronic state. For molecules, if we assume the Born-Oppenheimer approximation then within each electronic state the energy will vary as interatomic distances change and the molecules may also possess rotational energy. For a diatomic molecule the electronic energy level diagram is thus presented as a set of potential energy curves, as in Figure 4.1.

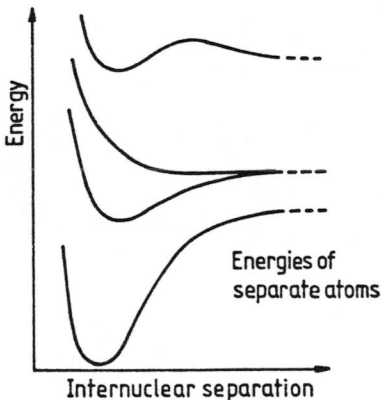

Figure 4.1 Electronic energy levels
of a diatomic molecule

A similar diagram for a polyatomic molecule would be very complex as hypersurfaces of more than two dimensions may be necessary to explain each electronic state. In practice, therefore, for polyatomic molecules, electronic energy level diagrams are presented as curves like those for diatomic molecules, the curve being a section through a hypersurface, or even just as a horizontal line as in atoms.

44

A. Electronic energy levels

Promoting an electron into an excited molecular orbital requires amounts of energy comparable to those needed in atomic excitation. In consequence the electronic absorption spectra come in the ultraviolet, visible and near infrared regions of the spectrum.

Much of the detail of an electronic transition can be predicted from the appropriate molecular orbital diagram as should be clear from one or two examples.

Hydrogen

The ground state configuration of H_2 is $1\sigma_g^2$. Excitation of an electron to the antibonding $1\sigma_u$ orbital can yield $^3\Sigma_u^+$ or $^1\Sigma_u^+$ excited states, depending on whether the electrons have parallel or opposed spins. Neither of these states is bound; that is to say there are no minima on the potential curves as shown in Figure 4.2. A ramification of this is that the allowed $^1\Sigma_u^+ - {}^1\Sigma_g^+$ has no fine structure and the spectrum is continuous. Indeed, this transition may be used as a source of 'white' light. The energy region is in the vacuum ultraviolet because $1s$ electrons on H or $1\sigma_g$ electrons on H_2 are tightly bound as a result of the influence of an unshielded nucleus.

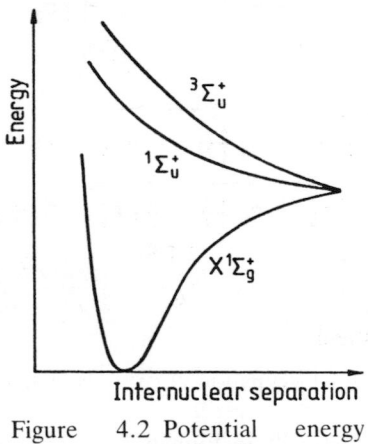

Figure 4.2 Potential energy curves for H_2

A common notation uses the prefix X to label the ground state, with excited states of the same multiplicity being A, B, C, etc., with those of an alternative multiplicity indicated by lower case letters, a, b, c, etc.

Oxygen

The higher orbital energy levels of O_2 are reproduced in Figure 4.3. In the case of O_2 with its unpaired π_g electrons we can go to higher electronic states without promoting electrons into the empty $3\sigma_u$ orbital. The two low-lying excited states

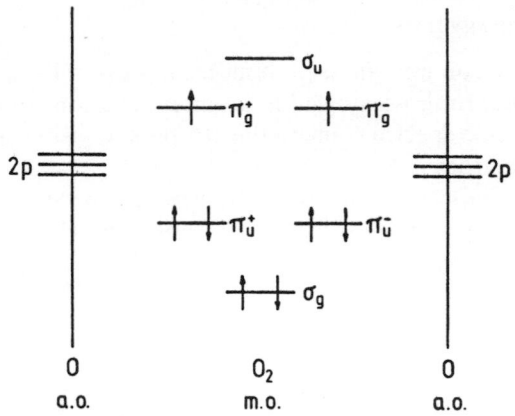

Figure 4.3 Molecular orbital diagram for the higher orbitals of O_2 in its ground electronic state

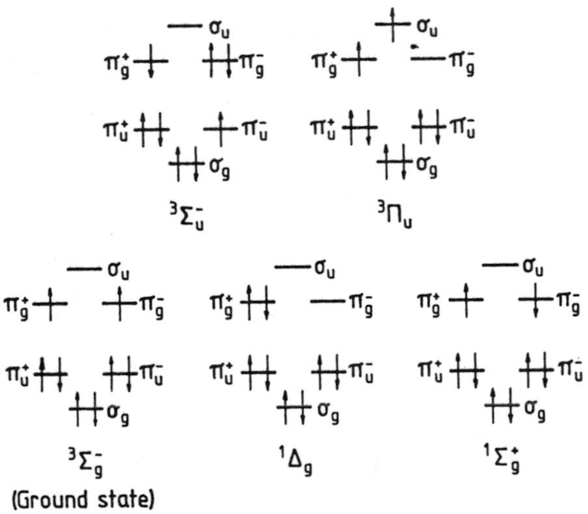

$^3\Sigma_u^-$ $^3\Pi_u$

$^3\Sigma_g^-$ $^1\Delta_g$ $^1\Sigma_g^+$

(Ground state)

Figure 4.4 Occupation of orbitals in states of O_2

are illustrated diagrammatically in Figure 4.4 together with the ground state and more highly excited states. The resulting potential curves are shown below. The selection rules forbid the $^1\Sigma_g^+ - {}^3\Sigma_g^-$ and $^1\Delta_g - {}^3\Sigma_g^-$ transitions but the $^3\Sigma_u^- - {}^3\Sigma_g^-$ transition is strongly allowed and is responsible for cutting out the shorter wavelength ultraviolet light in the earth's atmosphere. Note that the conventional habit in describing the electronic transitions of small molecules is to put the upper of the two states involved first.

B. Vibrational structure

Since we are dealing with electronic states which are represented by curves or surfaces rather than as lines, then an electronic transition appears on the

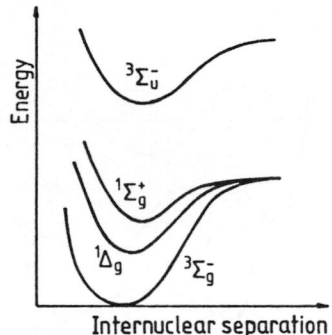

Figure 4.5 Potential curves for
low-lying states of O_2

spectrum of a molecule as far more than a single line. There may be extra structure due to the possible vibrations of the molecule and indeed its rotations. The vibrational energy of a given state gives rise to a set of levels which can to a large degree be treated independently. These levels are observed not only as structure in an electronic transition, but transitions between the vibrational levels are also observed in the infrared region of the spectrum.

The origin of vibrational energy levels is in the solution of the Schrödinger wave equation for a particle in the appropriate potential. If we have a square-well 'particle in a box' potential then quantized energy levels arise from the postulates that the allowed wavefunctions have to be well-behaved. This is summarized in Figure 4.6.

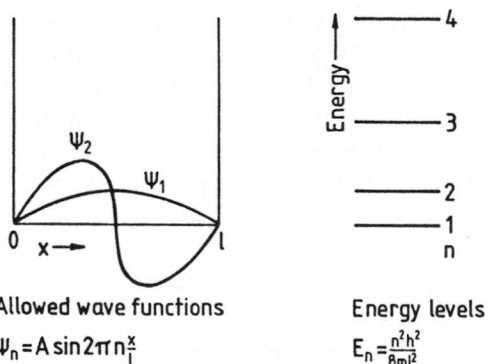

Allowed wave functions

$$\Psi_n = A \sin 2\pi n \frac{x}{l}$$

Energy levels

$$E_n = \frac{n^2 h^2}{8ml^2}$$

Figure 4.6 The wavefunctions and energies of a
particle in a box

Closer to the reality of a diatomic molecular potential energy curve is the harmonic oscillator potential. The assumption that a diatomic molecule can be represented by two weights on the end of a spring (the bond) is quite realistic, especially for the all-important region close to the minimum. Figure 4.7

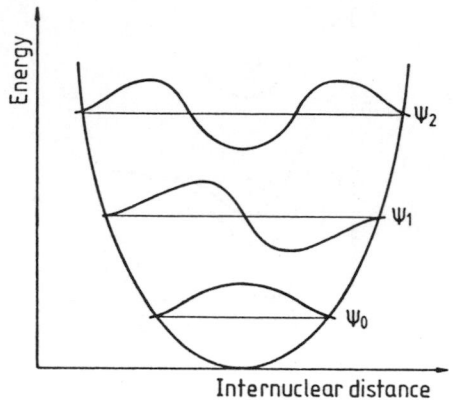

$$E=(v+\tfrac{1}{2})h\nu$$

Figure 4.7 Energy levels and wavefunctions of an harmonic oscillator

summarizes the wavefunctions and energy levels of the harmonic oscillator. The levels are labelled 0, 1, 2 from the bottom and even the lowest energy level involves some vibration. The energy of this vibration is called the zero-point energy and is equal to $\tfrac{1}{2}h\nu$. This is a direct consequence of the uncertainty principle. The wavefunctions, in common with those we have already encountered in the case of atoms, have an increasing number of nodes as the energy increases. It should also be noted that if the functions are inverted through the mid-point, then alternately as one goes up the series they are either odd or even in the sense that inversion involves a sign change or it does not.

A real diatomic potential curve, although being very similar to an harmonic oscillator at the bottom of the well, diverges at higher energies. The left-hand side increases in energy indefinitely, reflecting the increasing difficulty of forcing charges to get closer and closer. However, the right-hand side, large internuclear separation, flattens out at the point where the bond is broken. The result of this anharmonicity is to cause the energy levels to get closer together higher up the curve as in Figure 4.8, finally giving a continuum of levels.

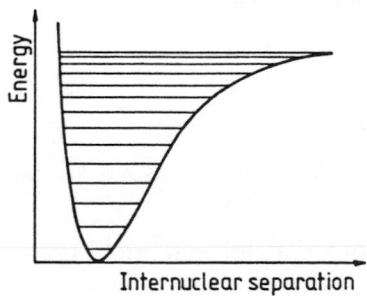

Figure 4.8 Potential curve for a
diatomic molecule

A mathematical expression which will fit this type of behaviour well is

$$E_v = (v + \tfrac{1}{2})\omega_e - (v + \tfrac{1}{2})^2 x_e \omega_e + (v + \tfrac{1}{2})^3 y_e \omega_e + \cdots,$$

where v is the vibrational quantum number, $\omega_e = h\nu$ and $x_e\omega_e$ is a small anharmonic correction. Further terms, $y_e\omega_e$ and beyond are all increasingly small and can be ignored unless very accurate work is required or unless values of v become very large, when $(v + \tfrac{1}{2})^2$ can become important.

C. Rotational structure

Molecules can rotate as well as undergo internal vibrations. The rate of vibration is, on the other hand, so rapid by comparison with the rotation that we can consider the body which rotates as an average structure over the vibrational motion. Taking the diatomic molecule as the simplest example, the vibration is so rapid that the effective rotating molecule just has the average internuclear separation time-averaged over the period of a vibration. To a first approximation we can consider the rotating diatomic molecule as a rigid rotating body as in Figure 4.9, with a fixed internuclear bond length during the period of a vibration.

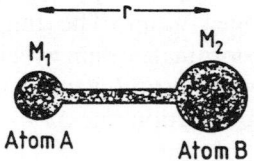

Figure 4.9 The rigid rotor

Just as vibrational energy levels can be considered separately from electronic levels, so rotational energy levels can be treated in isolation. These rotational levels will give fine structure imposed upon the vibrational structure of an electronic transition; they will give fine structure in a vibrational transition within one electronic state; and pure rotational transitions are also possible, involving transitions between rotational levels without changing either vibration or electronic states.

The actual form of the rotational levels can be found from the solution of the Schrödinger wave equation for a rigid rotor. The potential energy term V which has to be introduced is simply $V = 0$, since no potential energy is associated with rotation so long as the rotator remains rigid with a fixed bond length. In these circumstances for a diatomic molecule the moment of inertia is

$$I = \frac{m_1 m_2}{m_1 + m_2} \, r^2$$

or

$$I = \mu r^2, \qquad \mu \text{ being the reduced mass.}$$

50

The appropriate Schrödinger equation is thus

$$\nabla^2 \Psi + \frac{8\pi^2 \mu}{h^2} E\Psi = 0.$$

The comparatively simple solution of this equation which satisfies the constraints which are postulated for wavefunctions leads to the energies and wavefunctions depicted in Figure 4.10.

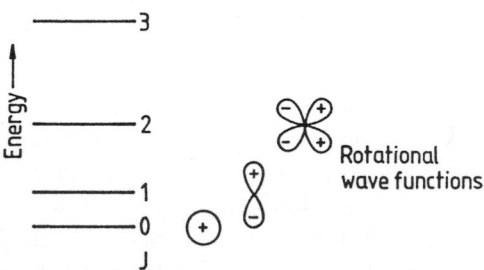

Figure 4.10 Rotational energy levels and wavefunctions for the rigid rotor

It is interesting to note that the rotational wavefunctions are none other than our old friends the spherical harmonic functions which appear in the solution of the Schrödinger equation for the hydrogen atom. The functions have a symmetry with respect to inversion at the coordinate origin which is alternately symmetric (no change of sign) and antisymmetric (change sign). Thus we should not be surprised to encounter later a selection rule $\Delta J = \pm 1$ rather like the atomic rule $\Delta L = \pm 1$.

The form of the energy levels of a rigid rotor as given above in Figure 4.10 can also be seen intuitively if we are prepared to accept a result which comes from a full treatment, namely that the angular momentum, P, of a rigid rotor is quantized in units of $h/2\pi$ in the manner:

$$P = \frac{h}{2\pi} \sqrt{J(J+1)} \ .$$

Now classically $P = I\omega$, with ω being the angular velocity, and the energy of a classical rotor is

$$E = \tfrac{1}{2}I\omega^2$$
$$= \tfrac{1}{2} \frac{(I\omega)^2}{I} = \frac{h^2 J(J+1)}{8\pi^2 I}$$
$$= \frac{h^2 J(J+1)}{8\pi^2 \mu r^2},$$

the result given above. This may be rewritten as

$$E = BJ(J+1),$$

with B incorporating the fundamental constants and the molecular properties of reduced mass and bond length.

A correction to this formula may be made by allowing for the fact that as a molecule rotates the bond may stretch slightly—centrifugal distortion. We need to alter the formula so that at high rotation speeds (large values of the rotational quantum number J) the bond length r can be increased. This would reduce the value of B which contains $1/r^2$ for large J and can be achieved by rewriting the energy level expression as

$$E = BJ(J + 1) - DJ^2(J + 1)^2,$$

where D is a small correction term.

Polyatomic molecules can be subdivided into several classes. Linear molecules like CO_2 have only a single moment of inertia and the above formulae apply except that I cannot be replaced by μr^2.

In general a body has three principal amounts of inertia, one about each perpendicular axis, usually designated as I_A, I_B and I_C. For the linear molecules $I_B = I_C$ and $I_A = 0$.

Molecules are designated as symmetric tops (e.g. CH_3Cl) if $I_B = I_C \neq I_A$ and $I_A \neq 0$. Their rotational energy levels are given by the formula

$$E = BJ(J + 1) + (A - B)K^2,$$

where B is equal to $h/8\pi^2 I_B$, A is equal to $h/8\pi^2 I_A$ and K is a second quantum number. Spherical tops have $I_A = I_B = I_C$, e.g. CCl_4 or CH_4.

Asymmetric tops have $I_A \neq I_B \neq I_C$ and have very complicated sets of energy levels which cannot be calculated from simple formulae.

D. The effects of fields

Many of the energy levels just described are degenerate, i.e. there may be several levels of identical energy. This degeneracy may be removed and the levels split by the imposition of magnetic or electric fields, paralleling in many respects the Zeeman and Stark effects familiar in atomic spectroscopy.

If we put a rotating molecule into a magnetic field then the angular momentum vector may only take up directions such that its component in the direction of the field is quantized. Thus, just as in the Zeeman effect in atoms, the rotational energy levels of a molecule which can be labelled by the quantum number J are $(2J + 1)$-fold degenerate. The example of the case where $J = 1$ is illustrated in Figure 4.11. This degeneracy will be of importance when we come to consider the relative intensities of spectral lines and the population of energy levels.

The removal of the degeneracy by field is important in the case of electric fields and the molecular Stark effect is used in pure rotational microwave specroscopy, both in the method of spectral detection and to measure dipole moments.

The different energy sublevels produced by magnetic and electric fields normally differ physically in the orientation of the spins of either electrons or nuclei. The different orientations of the former give rise to larger energy gaps

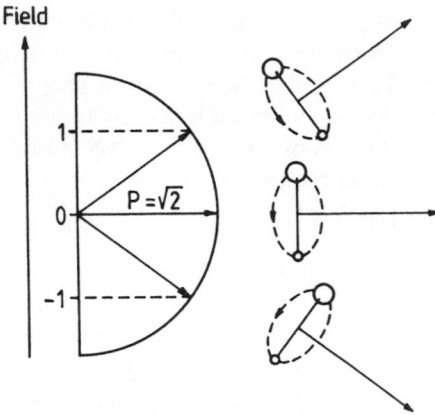

Figure 4.11 The orientations of the rotational angular momentum for $J = 1$

than the latter, although in both cases the magnitude of the splitting of formerly degenerate levels is a function of the applied field.

These sublevels can give rise to branches of spectroscopy in their own right with transitions among themselves or they may add hyperfine structure to rotational, rotation-vibrational or electronic transitions.

E. Spectroscopy

Spectroscopy is the measurement of the interaction of electromagnetic radiation with matter to cause changes in the population of any of the various energy levels of a molecule. The electromagnetic radiation which can be absorbed as a molecule goes from an energy level of lower energy to one of the higher energy, or emitted in the reverse process, has oscillating electric and magnetic field components. In most forms of spectroscopy it is the oscillating electric field of the radiation which is the more important.

Following Planck's relation, $E = h\nu$, it is clear that radiation of high frequency, ν, has high energy and short wavelength, λ, since velocity $c = \nu\lambda$.

In the subsequent chapters we will consider each of these types of transition in turn, starting at the long wavelength end of the spectrum. In this way the energy levels encountered in each chapter will serve as the fine structure for the type of spectroscopy met in the subsequent chapters, although spin resonance will be taken out of sequence.

To understand a branch of spectroscopy it is first necessary to understand the energy levels involved and then to know which transitions between them are allowed—that is to say the selection rules. A full understanding includes a third factor, the intensities and relative intensities of lines. The whole subject of intensity and the derivation of selection rules will be left for a single chapter after we have discussed some branches of spectroscopy because so many of the principles are common to every type of spectral transition.

F. Summary

Molecules have quantized rotational levels with a spacing which corresponds to transitions in the microwave region of the spectrum. Such transitions give rise to pure rotation spectra. In the infrared, vibrational transitions take place with rotations adding fine structure. Visible and ultraviolet light cause electronic transitions which are observed with vibrational energy changes superimposed and rotational transitions providing fine structure.

G. Problems

1. The vibrational energy levels of the diatomic molecule NaI lie at the following wavenumbers:

   $$129.28; 389.48; 651.88; 916.48 \text{ cm}^{-1}.$$

   Deduce the value of the constants ω_e and $x_e\omega_e$ and the zero point energy.
2. The molecule O_2 has a vibrational frequency of 1580 cm^{-1}, and a rotational constant B of 1.45 cm^{-1}. Which is the lowest rotational level of the $v = 0$ state to have a higher energy than the $J = 0$ level of the $v = 1$ state? What approximation is made in this calculation?
3. The vibration frequency of a $^{35}Cl_2$ molecule is 1.68×10^{13} Hz. Convert this into joules, and use the Boltzmann law to calculate the temperature at which 5 per cent of the molecules in a sample of Cl_2 are in the $v = 1$ state. (You may neglect the population in higher vibrational states : why?)
4. As well as an intramolecular potential which gives rise to the potential curve for a diatomic molecule such as H_2 there is also an intermolecular potential between H_2 molecules which has the same qualitative shape although the attractive energy is very much less. There is a zero-point energy for the H_2–H_2 potential as well as within H_2. Suggest differences in physical properties of liquid H_2 or D_2 which could arise from differences in intermolecular zero-point energy.
5. Suggest what effects the following types of radiation might have on molecules in food which is being cooked: microwaves; infrared radiation from a barbecue; radiowaves; ultraviolet light.

Chapter 5

Pure Rotation

The simplest spectra of small molecules are found in the microwave and far–infrared regions of the electromagnetic spectrum; they correspond to the molecules changing just their rotational energy.

A. Diatomic molecules

We have seen already that a diatomic molecule is capable of vibrating and rotating; the allowed energy levels for a typical molecule are shown in Figure 5.1. The difference in energy between the two vibrational levels is much larger

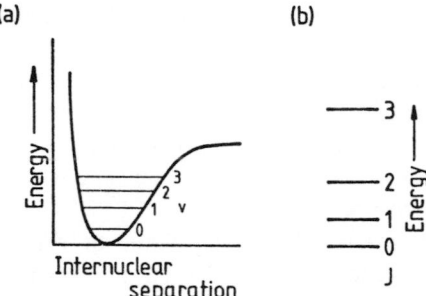

Figure 5.1 Vibrational (a) and rotational (b) energy levels of a diatomic molecule (not to same scale)

than that between two rotational levels, and in this chapter we are concerned with radiation of too low an energy to change the vibrational quantum number. The period of vibration of a diatomic molecule is much shorter than the period of rotation, and so for any particular rotational level we can regard the molecule as being a rigid body, with a time-averaged value of the bond length.

As seen in Chapter 4, the Schrödinger equation gives the energy levels of a rigid rotor as

$$E_{\mathrm{rot}} = BJ(J + 1) \tag{5.1}$$

where

$$B = \frac{h^2}{8\pi^2 \mu r^2} \tag{5.2}$$

With the aid of selection rules, it is possible to predict the appearance of the rotational spectrum. The first selection rule is that in order to absorb electromagnetic radiation a molecule must have a dipole moment. In a simple manner this can be understood by remembering that light can only interact with a molecule by means of its oscillating electric or magnetic field; for the more important electric dipole transition, this means that the molecule must either have a permanent electric dipole or the transition must involve the redistribution of electrons. As microwaves are insufficiently energetic to move an electron from one orbital to another, a permanent dipole is a prerequisite for a rotational spectrum. This means that all homonuclear diatomic molecules, such as H_2, N_2 and Cl_2, have no pure rotational spectrum.

For a heteronuclear molecule, the frequency of any transition will be given by the formula

$$h\nu = \Delta E = BJ'(J' + 1) - BJ''(J'' + 1),$$

where we have followed the standard practice of using a single prime for the upper state involved in the transition and a double prime for the lower state. The selection rule for pure rotational transitions is

$$\Delta J = \pm 1.$$

This selection rule means that the angular momentum of the molecule changes by one unit, which we can understand as an example of the law of conservation of angular momentum, the one unit being carried by the incoming or outgoing photon. Thus

$$J' = J'' + 1,$$

and

$$h\nu = \Delta E = B(J'' + 1)(J'' + 2) - BJ''(J'' + 1)$$
$$= 2B(J'' + 1) = 2BJ'.$$

Thus the allowed values of $h\nu$ are $2B$, $4B$, $6B$, etc., and the spectrum should consist of a series of equally spaced lines, as shown in Figure 5.2.

The value of B depends on the moment of inertia of that molecule and hence mainly on the reduced mass. The hydrogen halides, which have a very low reduced mass, absorb light in the far infrared region of the spectrum, from 20 to 100 cm^{-1}, whereas the molecules with a greater reduced mass, such as CO, absorb in the microwave region, around 1 cm^{-1}. Remember that the curious unit cm^{-1}, the number of waves that will fit in 1 cm, is only used as a measure of energy; its relationship to other units of energy is given in Figure 1.6.

Measurement of the separation of the lines in the rotational spectrum allows the bond length of the molecule to be determined; for HCl the separation is

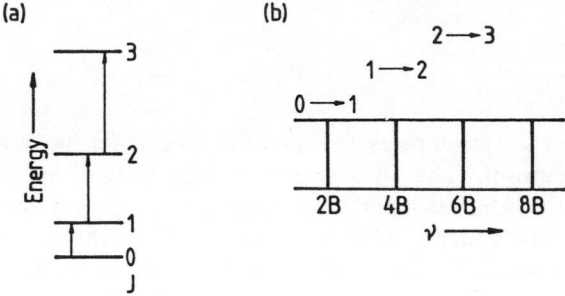

Figure 5.2 (a) Transitions between rotational energy
levels. (b) The spectrum

20.7 cm^{-1} and therefore B has the value of 10.35 cm^{-1}. We can
convert this to more useful units:

$$
\begin{aligned}
B &= 10.35 & \text{cm}^{-1} \\
&= 10.35 \times c & \text{Hz} \\
&= 10.35 \times 3 \times 10^{10} & \text{Hz} \\
&= 10.35 \times 3 \times 10^{10} \times h & \text{J} \\
&= 10.35 \times 3 \times 10^{10} \times 6.63 \times 10^{-34} & \text{J} \\
&= 2.06 \times 10^{-22} & \text{J}
\end{aligned}
$$

Now, for HCl,

$$
\begin{aligned}
m_1 &= 1/(6 \times 10^{-23}) & \text{g} \\
m_2 &= 35/(6 \times 10^{-23}) & \text{g}
\end{aligned}
$$

Therefore,

$$
\begin{aligned}
\mu &= 35/(36 \times 6 \times 10^{-23}) & \text{g} \\
&= 1.62 \times 10^{-24} & \text{g} \\
&= 1.62 \times 10^{-27} & \text{kg}
\end{aligned}
$$

Therefore, from equation (5.2):

$$
r^2 = \frac{(6.63 \times 10^{-34})^2}{8 \times 3.142^2 \times 1.62 \times 10^{-27} \times 2.06 \times 10^{-22}} \ \text{m}^2
$$

$$
\begin{aligned}
r &= 1.29 \times 10^{-10} \text{m} \\
&= 0.129 \text{ nm}
\end{aligned}
$$

Careful measurements of the spectra of some diatomic molecules show that
the observed lines are not exactly equally spaced, but in fact converge slowly.
This is because a molecule is not strictly a rigid rotor; the molecule can stretch
and at high rotational speed the internuclear separation increases. This can be
taken into account by modifying the equation, giving a more accurate
equation for the rotational levels of a diatomic molecule:

$$
E_{\text{rot}} = BJ(J + 1) - DJ^2(J + 1)^2.
$$

Here D is very much smaller than B; for HCl, B has the value $10.35 \, \text{cm}^{-1}$ and $D = 0.0004 \, \text{cm}^{-1}$.

The magnitude of D will depend on the ease with which the bond can stretch; if the bond has a small vibrational frequency ω then the potential curve will be shallow and D large. It can be shown that the relationship between D and ω is

$$D = \frac{4B^3}{\omega^2} \, .$$

Thus it is sometimes possible to measure the vibration frequency of a diatomic molecule from the pure rotation spectrum, although this is often rather inaccurate as D is so small.

B.   Polyatomic molecules

Linear molecules

For linear polyatomic molecules, the energy levels and formulae are identical to those encountered for diatomic molecules. However, now the measured value of B yields only a value of the moment of inertia, and not a value of the bond lengths,

$$E_{\text{rot}} = \frac{h^2}{8\pi^2 I} \, J(J + 1).$$

If we consider a simple example, OCS, we can see why bond lengths do not follow immediately from the spectrum. (It might be thought that CO_2 would be a simpler example, but that molecule is symmetrical and therefore has no dipole moment.) The spectrum of OCS has been extensively studied, and from the separation of lines the B value has been calculated and hence the moment of inertia calculated as $1.38 \times 10^{-45} \, \text{kg m}^2$. It is clear from this one piece of information that it is not possible to calculate two unknown bond lengths. This problem can only be overcome if the experiment is repeated using isotopic substitution. The first experiment used the common $O^{16}C^{12}S^{32}$; the second experiment might use $O^{16}C^{12}S^{34}$. The important point is that the use of a different sulphur isotope does not change the bond lengths in the molecule, for the bond lengths are determined only by the electronic structure and not by the mass of the nuclei. However, $O^{16}C^{12}S^{34}$ has a different moment of inertia, and this is measured from the pure rotation spectrum. We now have two equations in two unknowns, and these can be solved giving the bond lengths:

C = O $1.165 \times 10^{-10} \, \text{m}$

C = S $1.558 \times 10^{-10} \, \text{m}$.

Spherical tops

Spherical tops are molecules which have all three moments of inertia identical, $I_A = I_B = I_C$. Simple examples are CCl_4 and SF_6. They also have no permanent dipole moment and accordingly they do not give rise to pure rotational spectra.

Symmetric tops

Symmetric tops are molecules which have two equal moments of inertia about perpendicular axes and the third moment of inertia non-zero; a simple example is CH_3Cl. The rotational energy of a symmetric top is given by

$$E_{rot} = BJ(J + 1) + (A - B)K^2,$$

where J is the quantum number describing the total angular momentum and K the quantum number describing the angular momentum about the symmetry axis. A and B are the rotational constants corresponding to the moments of inertia I_A and I_B; here $I_A \neq I_B = I_C$. The selection rules for pure rotational transitions are

$$\Delta J = \pm 1$$
$$\Delta K = 0$$

and hence

$$h\nu = \Delta E = 2BJ',$$

just as for diatomic molecules.

Pure rotational spectroscopy thus only reveals information about one moment of inertia for a symmetric top; again, further information is obtained by isotopic substitution.

Asymmetric tops

Asymmetric tops are molecules with three different, non-zero moments of inertia. They give rise to line spectra in the microwave region, but their energy levels cannot be expressed in any simple form. In a number of cases the spectra have been analysed, but the analysis is beyond the scope of this book. These analyses can give accurate information not only on bond lengths and angles but in molecules such as methanol, CH_3OH, on the energy barrier to the rotation of the $-CH_3$ group.

C. The Stark effect

When a molecule is placed in an electric field, the energy of the rotational levels is shifted, as the electric field interacts with the molecular dipole moment. This effect is called the Stark effect. Just as in the atomic case, not only is the quantum number J defined, but also M_J, the component of J in the direction of the electric field. M_J can take all integral values from $+J$ to $-J$.

In a linear molecule, the dipole moment and the angular momentum are necessarily perpendicular to each other, and there is therefore no first-order Stark effect. However, there is a smaller second-order effect, and the energy change produced by the electric field is given by

$$\Delta E = \frac{4\pi^2 I\mu^2}{h^2} \cdot \frac{J(J + 1) - 3M^2}{J(J + 1)(2J - 1)(2J + 3)} \cdot E^2,$$

where E is the electric field strength in volts per metre and μ is the dipole moment in metrecoulombs. This produces an energy level diagram as shown in Figure 5.3.

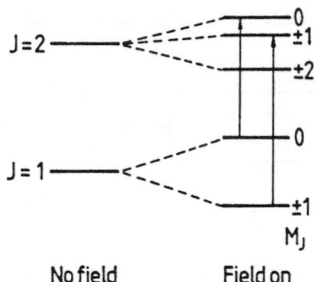

Figure 5.3 The splitting of the $J = 2 \leftarrow J = 1$ transition in an electric field

The selection rules $\Delta J = \pm 1$, $\Delta M_J = 0$ apply, and the line is split by the electric field into two. The magnitude of the splitting is easily measured, and from this splitting the dipole moment of the molecule may be calculated. This is an important and accurate method of measuring dipole moments; often the electric field strength is the most difficult quantity to measure, and this can be done by using the Stark effect of a molecule of known dipole moment, such as OCS. Another valuable feature of the Stark effect is that it allows unambiguous determination of J values; the number of components into which a line is split depends on the value of J, which is otherwise difficult to determine if the first few lines of a series are difficult to observe, either because of their low intensity or for experimental reasons.

Symmetric top molecules have a first-order Stark effect, so in general rather small electric field strengths are needed to observe the effect. The energy change produced by the electric field is given by

$$\Delta E = \frac{\mu K M}{J(J + 1)} E,$$

so again it is possible to use the Stark effect to measure the dipole moment.

60

D. Experimental microwave spectroscopy

Microwave spectra are studied in absorption; however, the properties of microwaves are such that there are some important differences between the microwave experiment and the more familiar experiments using visible or near infrared radiation. The apparatus used is shown in block diagram form in Figure 5.4.

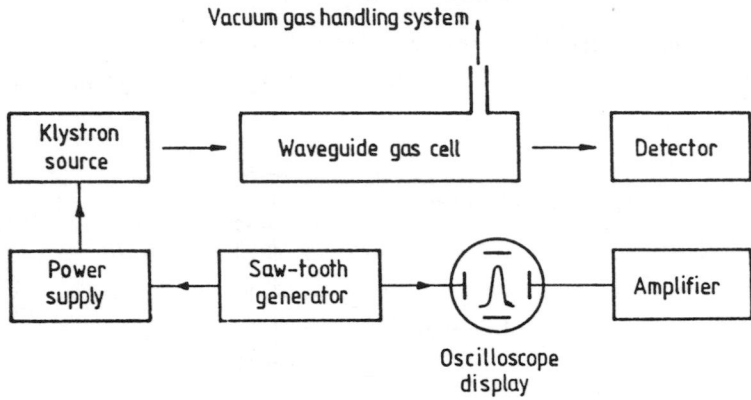

Figure 5.4 Experimental arrangement for a simple microwave spectro-meter

A convenient source of microwaves is the Klystron valve; this produces a highly monochromatic beam of microwaves, whose frequency may be controlled by varying the applied voltage. The radiation is passed through a hollow metal pipe called a waveguide, and passes through the gas sample. The gas is held in a waveguide with mica windows. After passing through the sample, the radiation is conveniently detected by a crystal diode. The spectrum can conveniently be displayed on an oscilloscope.

The experiment may be modified somewhat when Stark-modulated detection is employed. The experimental arrangement is shown in Figure 5.5. There is an insulated strip in the centre of the gas sample, and to this is applied a high voltage at a frequency of about 100 kHz. The voltage is applied such that for half the time there is no electric field in the sample chamber and for the other half there is a large electric field. This electric field produces a Stark effect, which splits and broadens the absorption spectrum; thus when the Klystron is emitting the correct frequency for molecular absorption, a 100 kHz signal appears at the detector.

As always, any spectroscopic measurement is made by comparing the frequencies of lines with standard lines whose frequencies are accurately known; quartz crystal controlled oscillators may be used for this purpose, but radio station signals are commonly used.

One of the most remarkable features of microwave spectroscopy is the high accuracy of the data obtained from it. This is partly due to the high sensitivity of

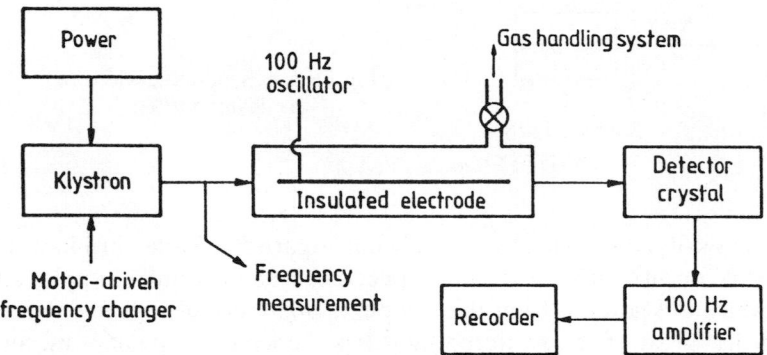

Figure 5.5 Stark modulation gas microwave spectrometer

the apparatus used, but this is only important because the linewidth of microwave spectra is extremely small. No spectral line is infinitely sharp, but in fact spreads over a small range of frequencies. We shall study the factors which influence linewidths in greater detail in Chapter 8, but for the moment we can note two factors which contribute to the small linewidth. Firstly, the gas pressure in the microwave experiment is normally very low and so pressure broadening is not an important factor. Secondly, the experiment may be conducted at room temperature (or even below) and so broadening by the Doppler effect is negligible.

In favourable cases, frequencies of microwave lines have been observed to seven significant figures, at which level the accuracy to which the speed of light is known is an important factor governing the accuracy to which bond lengths can be determined. In diatomic molecules, bond lengths can usually be determined to 0.0001 Å, and slightly less accurately in the case of simple polyatomic molecules.

With such accurate data available, it is necessary to be rather careful in stating exactly what we mean by a bond length. For a diatomic molecule in its ground vibrational state, the bond length r_0, is strictly given by

$$r_0 = \left(\frac{1}{r^2}\right)_{\text{average}}^{-1/2},$$

where the average is taken over the zero-point vibrational motion. This is not the same as r_e, the internuclear separation corresponding to minimum energy, but the correction can be made fairly easily and is typically less than 1 per cent. If r_e values are compared for two molecules differing only by one isotope, then they are found to agree to a high accuracy.

E. Pure rotational Raman spectra

We have seen that it is not possible to study the rotational levels of molecules which have no dipole moment by microwave spectroscopy. However, in most cases it is possible to obtain such information using Raman spectroscopy.

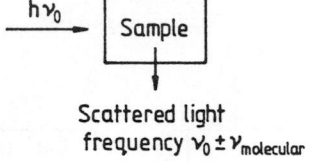

Figure 5.6 Schematic view of
the Raman effect

The basis of the Raman effect is shown in Figure 5.6. Monochromatic light in the visible or ultraviolet part of the spectrum is shone onto a sample, and the light which is scattered from the sample is analysed in a spectrometer. It is found that most of the scattered light has the same frequency, v_0, and this phenomenon is known as Rayleigh scattering. In addition, however, there are also components of the light with frequencies $v_0 - v_m$ and $v_0 + v_m$, where v_m is a frequency characteristic of the molecule. Light with frequency $v_0 - v_m$ is called Stokes radiation, the photons having lost some energy to the scattering molecule. Light with frequency $v_0 + v_m$ is called anti-Stokes radiation and here the scattered photons have picked up some of the quantized rotational energy from the molecule.

The selection rules for Raman scattering differ from those for the absorption of microwaves; the molecule must possess an anisotropic polarizability. The polarizability α is defined by the equation $\mu = \alpha E$, where μ is the dipole moment induced by an electric field of strength E. An example of such a molecule would be H_2, where it is easier to displace electrons along the $H-H$ bond than perpendicular to it. The only class of molecule which does not possess anisotropic polarizability is the spherical top.

The selection rule for Raman spectroscopy is

$$\Delta J = \pm 2,$$

where $\Delta J = 2$ for Stokes radiation and $\Delta J = -2$ for anti-Stokes radiation. Therefore,

$$h\nu = \Delta E = B(J + 2)(J + 3) - BJ(J + 1)$$
$$= 4BJ + 6B.$$

The Raman spectrum of a diatomic molecule therefore consists of a series of equally spaced lines, with spacing $4B$, except that the spacing between the Rayleigh line and the first line is $6B$. This applies to both homonuclear and heteronuclear diatomics, and is shown in Figure 5.7.

F.   Experimental Raman spectroscopy

The key experimental component of a Raman spectroscopy experiment is an intense source of monochromatic light. The intensity of the source is important, as the intensity of the Raman scattered light is very low indeed. For many years lines of the spectrum of atomic mercury were used, a typical experimental arrangement being shown in Figure 5.8. In recent years great

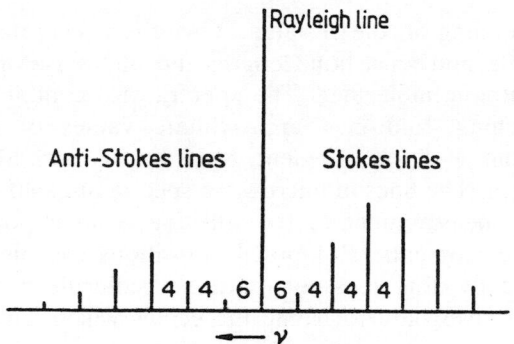

Figure 5.7 The Raman spectrum of a diatomic molecule

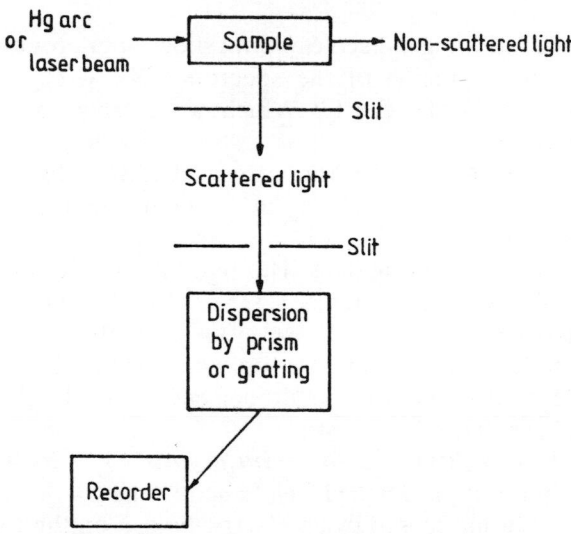

Figure 5.8 The experimental arrangement for Raman
spectroscopy

advances have been made in Raman spectroscopy with the use of lasers as light
sources. These are extremely monochromatic and have very high intensity.

The Raman effect permits the study of pure rotational levels of molecules
which do not have the dipole moment required for microwave spectroscopy. In
this sense it complements microwave spectroscopy, and in the next chapter we
shall see that Raman spectra can also complement vibrational spectra in a
similar way.

G.   Summary

Absorption of microwave radiation causes molecules to change their rotational
energy alone. Most molecules give line spectra in the microwave region, with

the spacing depending on the B value. This gives accurate information on moments of inertia, and hence bond lengths and angles; isotopic substitution is needed in polyatomic molecules. The spectra also contain information on molecular stretching, and give approximate values of bond strengths. Molecules without a dipole moment, such as N_2 and SF_6, do not give microwave spectra. The lines in microwave spectra are split when an electric field is applied; measurement of the splitting is an important method of measuring dipole moments. Rotational transitions can also be studied by Raman spectroscopy; this does not require a molecule to possess a dipole moment, and so gives data for some molecules which have no microwave spectrum.

H. Problems

1. Which of the following molecules would show a pure rotational spectrum in the microwave region of the spectrum: N_2, C_2H_4, CH_3OH, HD, CCl_4, CS_2, SO_2, NH_3, $BeCl_2$? Which would show a pure rotational Raman spectrum?

2. Rotational transitions for HCl are observed at 83.3, 104.1, 124.7, 145.4, 165.9, 186.2, 206.6 and 226.9 cm^{-1}. Find the moment of inertia and bond length of the molecule.

3. Comment on the following data. The first lines in the rotational Raman spectrum of $^{12}C^{16}O$ lie at ±11.535, ±19.225 and ±26.915 cm^{-1} from the exciting line. ($^{12}C = 12.000$, $^{16}O = 15.995$.)

4. The NH_3 molecule is a symmetric top with bond length 10.12 nm and a bond angle of 107°. Find its rotational energy levels. [Note: Moments of inertia, $I_A = 2m_1R^2(1 - \cos\theta)$
$$I_B = m_1R^2(1 - \cos\theta) + (m_1m_2/m)R^2(1 + 2\cos\theta)$$
with $m = 3m_1 + m_2$ and θ the H–N–H bond angle].

5. Predict the splitting caused by an electric field E on the rotational levels $J = 2$ and $J = 3$ of a diatomic molecule. Using the selection rules $\Delta J = \pm1$, $\Delta M_J = 0$ show what effect an electric field would have on the appearance of the $J = 3 \rightarrow J = 2$ absorption line. How can this be used to confirm assignments of J values in rotational spectra?

Chapter 6

Rotation–Vibration

Transitions between the vibrational energy levels of molecules appear in the infrared region of the electromagnetic spectrum. These transitions are almost always accompanied by a change in the rotational quantum number also, and the spectra are therefore called rotation–vibration spectra.

Once again it is convenient to discuss molecules in classes based on their overall structure.

A. Diatomic molecules

In Chapter 4 we saw that the simplest model for the vibration of a molecule is that of the harmonic oscillator. The Schrödinger equation can be solved exactly for this case, and the potential energy curve and vibrational energy levels are shown in Figure 6.1. Here the energy levels are equally spaced.

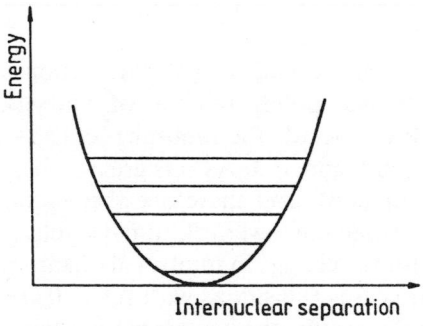

Figure 6.1 Energy levels of the harmonic oscillator

A more realistic model for a molecule is the anharmonic oscillator, with the potential curve becoming steeper at short internuclear distance and flattening off as the molecule dissociates. The vibrational energy levels of the anharmonic oscillator are rather similar to those of the harmonic oscillator and are shown in Figure 6.2. Two points are important here; firstly, the spacing between the energy levels decreases as v, the vibrational quantum number, increases and,

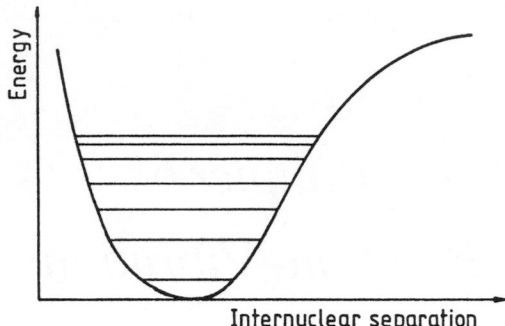

Figure 6.2 The lower vibrational energy levels
of a diatomic molecule

secondly, the mean internuclear spacing increases with v, and therefore B, the rotational constant, decreases with v.

One feature of vibrational spectra differs significantly from that of rotational spectra. At normal temperatures, it is common for molecules to occupy many rotational levels, and therefore rotational spectra consist of many lines, each corresponding to a different initial rotational state. However, in most cases only the lowest vibrational state is significantly populated, and therefore we shall be mainly concerned with transitions from the $v = 0$ state. This point will be discussed in more detail in Chapter 8.

Just as for pure rotation spectra, a diatomic molecule must have a dipole moment if an absorption spectrum is to be observed. Then, for an harmonic oscillator, the selection rule

$$\Delta v = \pm 1$$

applies, and so we are concerned with the transition $v = 0 \rightarrow v = 1$. (The selection rule $\Delta v = 0$ also applies, but this of course gives rise to the pure rotation spectrum.) If we include the rotational energy, then the energy level diagram is that shown in Figure 6.3. As r_1 is greater than r_0 for an anharmonic oscillator, B' is less than B'', and therefore the upper rotational levels are slightly closer together than the lower rotational levels.

The selection rule for the change in rotational quantum number is $\Delta J = \pm 1$; however, in some rare cases transitions with $\Delta J = 0$ are allowed. Remember that photons possess one unit of angular momentum, and therefore in an absorption experiment the rotational angular momentum must change by one unit. The exception to this rule occurs where there is another source of angular momentum in the molecule, spin or orbital angular momentum. This occurs when the electronic ground state of the molecule is not $^1\Sigma$ (e.g. $^2\Pi$, $^1\Delta$); NO is probably the only common molecule in this category. For NO, transitions with $\Delta J = 0$ are observed.

Spectral lines for which ΔJ is -1, 0 and $+1$ are respectively known as P, Q and R lines. They are labelled by the J value of the lower state; Figure 6.3 shows

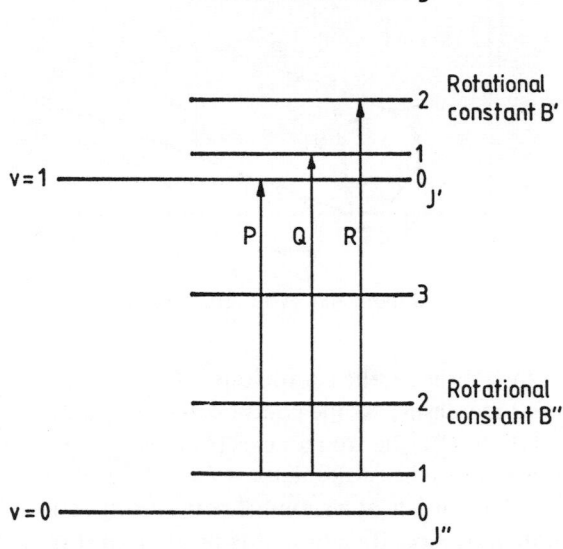

Figure 6.3 Vibration–rotation transitions

the transitions P(1), Q(1) and R(1). The set of lines P(1), P(2), P(3), … are known as the P branch.

The absorption frequency of a line is given by the general equation

$$h\nu = \Delta E = h\nu_{vib} + B'J'(J' + 1) - B''J''(J'' + 1)$$

if we neglect the centrifugal distortion terms discussed in Chapter 5. Hence we can obtain formulae for the P, Q and R branches of the spectrum:

$$h\nu_P = h\nu_{vib} + B'(J - 1)J - B''J(J + 1)$$
$$= h\nu_{vib} + \underset{-ve}{(B' - B'')}J^2 - \underset{-ve}{(B' + B'')}J$$

$$h\nu_Q = h\nu_{vib} + B'J(J + 1) - B''J(J + 1)$$
$$= h\nu_{vib} + \underset{-ve}{(B' - B'')}J^2 - \underset{+ve}{(B' - B'')}J$$

$$h\nu_R = h\nu_{vib} + B'(J + 1)(J + 2) - B''J(J + 1)$$
$$= h\nu_{vib} + 2B' + \underset{-ve}{(B' - B'')}J^2 + \underset{+ve}{(3B' - B'')}J.$$

Here the signs of the coefficients of J^2 and J have been shown, as they allow us to predict the general positions of the lines in each of the three types of branch. This is often shown in a Fortrat diagram.

B. Fortrat diagrams and band structure

A Fortrat diagram is a convenient way of plotting the functions which we have just obtained for P, Q and R branches; a typical diagram is shown in Figure 6.4.

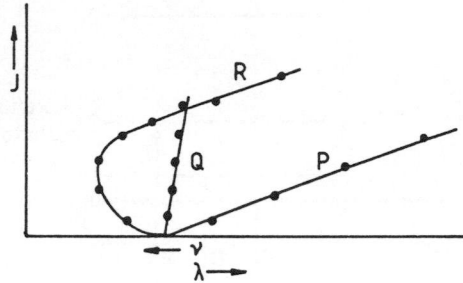

Figure 6.4 A Fortrat diagram

In the case of the P branch, the coefficients of both J and J^2 are negative, and so ν decreases as J increases. As the coefficient of J, $(B' + B'')$, is much larger than that of J^2, $(B' - B'')$, the frequencies progress almost linearly with J.

For the Q branch, both the coefficients of J and J^2 are very small, and so if a Q branch is observed it consists of a series of very closely spaced lines.

For the R branch, the coefficient of J is positive, but that of J^2 is negative, though much smaller. For low J values, the frequencies of the lines increase, but to higher values of J, the lines become closer together and eventually successive lines show a decrease in frequency. The band is then said to show a head.

For most practical cases, the difference between B' and B'' is very small, and as the intensity in the R branch decreases to zero after about $R(10)$, it is unusual to observe a band head. A typical absorption spectrum is shown in Figure 6.5. Notice that there is no Q branch.

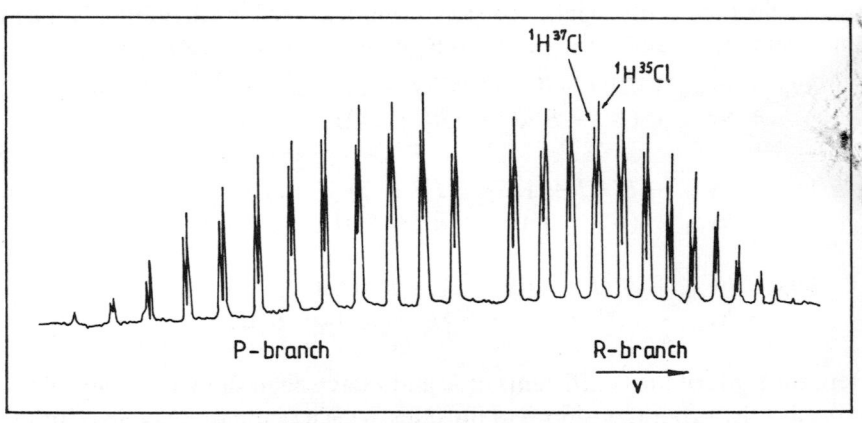

Figure 6.5 The vibration–rotation spectrum of HCl

C. Combination differences

It would be possible to extract values of B' and B'' by fitting observed frequencies of spectral lines to the formula given above. In practice it is more convenient, however, to consider the differences between the frequencies of pairs of lines which have a common upper or lower state. Then these frequency differences, or combination differences, Δ_2F, as they are known, depend only on the rotation constants of one level. This is shown in Figure 6.6.

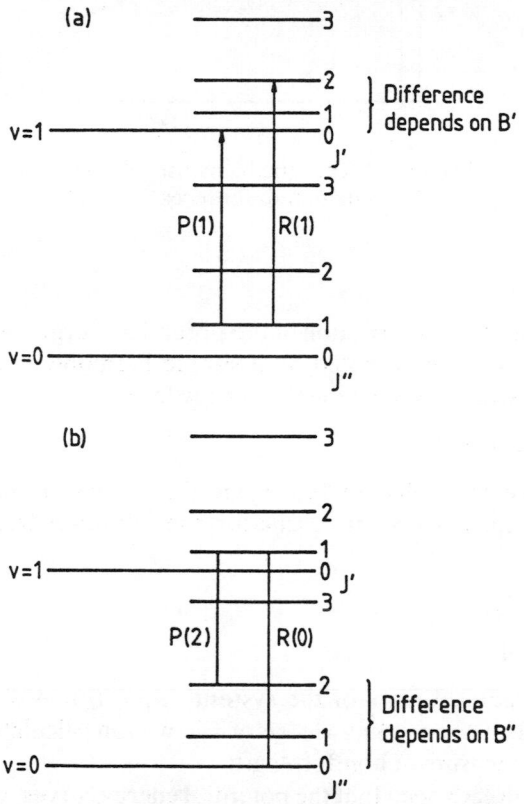

Figure 6.6 (a) Upper and (b) lower state combination differences

If we include the centrifugal distortion terms, then the full general formulae for combination differences of the upper and lower states are easily shown to be

$$\Delta_2F'' = R(J - 1) - P(J + 1) = 4B''(J + \tfrac{1}{2}) - 8D''(J + \tfrac{1}{2})^3$$

and

$$\Delta_2F' = R(J) - P(J) = 4B'(J + \tfrac{1}{2}) - 8D'(J + \tfrac{1}{2})^3.$$

Thus a graph of $\Delta_2F/(J + \tfrac{1}{2})$ versus $(J + \tfrac{1}{2})^2$ gives the value of B from the intercept and D from the gradient, as shown in Figure 6.7. From the rotational constants B_0 and B_1 it is possible to obtain values of r_0 and r_1, the internuclear spacing for the vibrational levels $v = 0$ and $v = 1$, using the formula

$$B = \frac{h^2}{8\pi^2\mu r^2}.$$

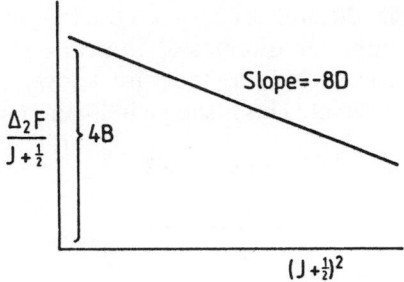

Figure 6.7 The graphical use of combination differences

D. The term ν_{vib}

We have seen that the lower portion of the potential energy curve of a diatomic molecule can be well represented by a simple harmonic oscillator, and the Schrödinger equation solved giving the energy levels

$$E_{vib} = h\nu_{vib} (v + \tfrac{1}{2}).$$

We can measure the value of ν_{vib} from the observed rotation–vibration spectrum. Now for a harmonic oscillator the vibration frequency is given classically by

$$\nu = \frac{1}{2\pi} \sqrt{\frac{k}{\mu}} \, ,$$

where μ is the reduced mass of the system, $m_1 m_1/(m_1 + m_2)$, and k is the force constant. Thus from a knowledge of ν_{vib} we can calculate the value of k, which is a rough measure of bond strength.

We have also already seen that the potential energy curves of the isotopically related molecules, such as H_2 HD and D_2, are identical, as the forces holding the molecule together are independent of the nuclear masses. In other words, they have the same force constant, and thus we can use the expression for ν_{vib} to calculate the relationship between the vibrational energy levels. These become closer together for heavier isotopes, as μ appears in the denominator of this expression; this is shown in Figure 6.8.

E. Anharmonicity

We know that the harmonic oscillator is not a very good model for the vibration of a diatomic molecule, particularly for high values of v. An empirical correction is usually made to allow for deviations from this simple model, the energy levels being rewritten as

$$E_{vib} = h\nu(v + \tfrac{1}{2}) - h\nu x(v + \tfrac{1}{2})^2$$

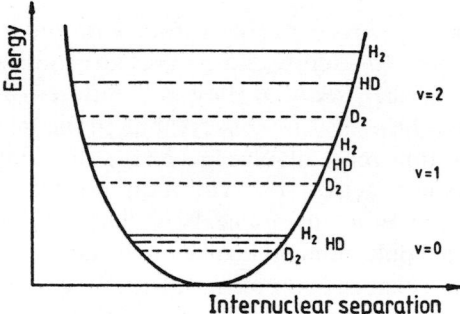

Figure 6.8 Vibrational levels for different
isotopic H_2 molecules

or

$$E_{vib} = \omega_e(v + \frac{1}{2}) - x_e w_e(v + \frac{1}{2})^2,$$

where x is called the anharmonicity constant. It is possible to include higher terms but they are usually negligible.

One important result of this anharmonicity is that it is no longer true that the selection rule $\Delta v = \pm 1$ applies; it is possible to observe transitions with $\Delta v = 2$, $\Delta v = 3$, etc., although in practice they have very low intensities. The full spectrum of a diatomic molecule thus consists of a band corresponding to the $v = 0 \rightarrow v = 1$ transition, called the fundamental, and much weaker bands, at frequencies roughly 2, 3, ... times the fundamental frequency, called overtones, corresponding to the $v = 0 \rightarrow v = 2$, $v = 0 \rightarrow v = 3$, etc., transitions.

It is not possible to calculate the values of both v and x by measuring just the fundamental band, but if measurements on the overtones are included, then both v and x may be calculated, and hence more precise information on the shape of the potential curve obtained.

F. Linear polyatomic molecules

The vibration of a polyatomic molecule is complicated, but it can be considered as the sum of a number of simple vibrations. The actual number of such simple vibrations can be obtained from the number of atoms in the molecule. If the molecule has N atoms, then it will have $3N$ degrees of freedom, and these correspond to the molecule moving, rotating and vibrating. Three of these degrees of freedom correspond to translation of the molecule along the three Cartesian axes, and three more correspond to rotation about these three axes. Of course, linear molecules can only rotate about two axes. Thus for non-linear molecules there are $(3N - 6)$ normal modes of vibration and for linear molecules $(3N - 5)$ normal modes.

Just as in microwave spectroscopy, both absorption and Raman experiments are possible, but it is now frequently useful to perform both experiments on the same molecule, as we shall see that they give different information. For a particular vibration to be active, i.e. observable, in the infrared region of the spectrum, that vibration must involve a change in dipole moment. For a vibration to be Raman active, the vibration must involve a change in polarizability. It is important to notice here that it is not necessary for a molecule to possess a dipole moment in order to be infrared active, merely that its vibration involves a change in dipole moment. To emphasize this point, we shall consider CO_2 as our example of a linear molecule.

G. The carbon dioxide molecule

Carbon dioxide is a linear molecule with three atoms, and therefore it has $(3 \times 3 - 5) = 4$ normal modes of vibration. These are illustrated in Figure 6.9. The figure also gives the standard labelling of these vibrations, v_1, v_2 and v_3, and indicates whether the vibrations involve a change in dipole moment, μ, or polarizability, α; v_2 and v_3 are also labelled $\parallel$ or $\perp$, to indicate whether the vibration is parallel or perpendicular to the molecular axis. Notice that v_2 really corresponds to two different vibrations, one with the atoms moving in the plane of the paper and the other with them moving into and out of the paper. These two have the same energy and frequency.

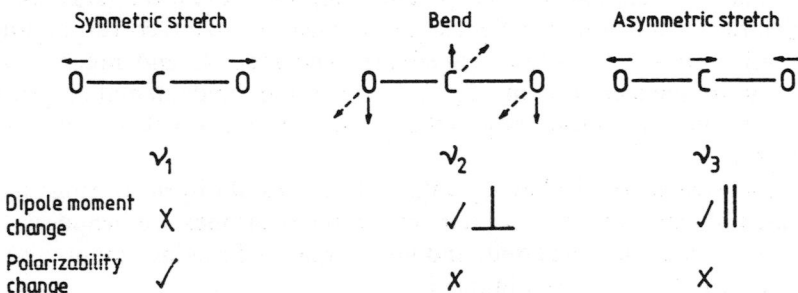

Figure 6.9 The vibrational modes of CO_2

The infrared region of the spectrum CO_2 should therefore show absorption at two distinct vibrational frequencies, v_2 and v_3, each with a superimposed rotational fine structure. The Raman spectrum will show just one vibrational frequency, v_1, again with rotational fine structure. Thus the infrared and Raman spectra will reveal different information about vibrations of the CO_2 molecule.

H. Rotational fine structure

The formulae for the rotational energy levels of linear polyatomic molecules are exactly the same as those of diatomic molecules, but the selection rules for infrared spectra are not quite so simple. For parallel vibrations, $\Delta J = \pm 1$, whereas for perpendicular vibrations $\Delta J = 0, \pm 1$. Thus for parallel vibrations P and R branches are observed, whereas for perpendicular vibrations P, Q, and R branches are observed. The general appearance of those spectra is given in Figure 6.10.

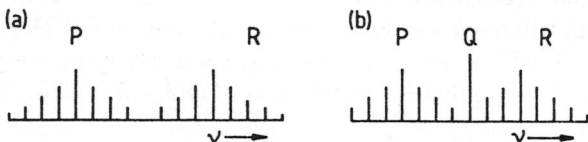

Figure 6.10 Rotational structure of (a) a parallel band, (b) a perpendicular band

For Raman spectroscopy, the selection rule is $\Delta J = 0, \pm 2$, and so O, Q and S branches ($\Delta J = -2$, 0 and +2 respectively) are observed. The general appearance of the spectrum is therefore as given in Figure 6.11, where the spacing of the rotational structure is twice that of the infrared spectrum.

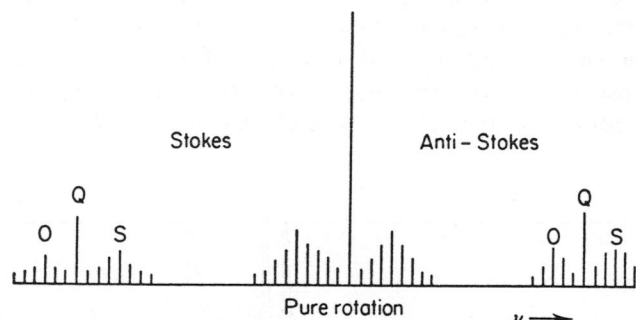

Figure 6.11 Schematic rotational–vibrational Raman spectrum

I. Assigning normal modes

In the case of CO_2 it is relatively straightforward to assign the various vibrational frequencies, using the infrared and Raman spectra. In more complicated cases, however, where there may be more atoms or where the molecule may not be linear, some tricky detective work may be required. Some of the features commonly used to assist in assigning vibrational spectra are:

(a) The mutual exclusion rule. When a molecule possesses a centre of symmetry, all vibrational modes which are infrared active are Raman inactive, and vice versa (CO_2 is an example of this rule).

74

(b) Bending frequencies are generally rather lower than stretching frequencies. Thus in CO_2 the bending frequency (v_2) is 667 cm^{-1}, whereas the stretching frequencies (v_1 and v_3) are 1340 cm^{-1} and 2349 cm^{-1}.

(c) Isotopic substitution may confirm ideas on assignment by revealing which vibrational frequencies are shifted. Thus in CH_3Cl, the C$-$Cl stretching frequency will move to a lower frequency than for Cl^{37} than for Cl^{35}.

(d) The rotational fine structure often assists in assigning normal modes, as the selection rules depend on the class of molecule and on the vibrational type.

(e) In Raman spectroscopy, if the incident light is polarized, then a totally symmetric vibration will produce polarized scattered light. This is not true of other vibrational modes, and measuring the degree of polarization can assist greatly in assigning vibrations. This has become particularly important with the advent of lasers as light sources.

J. Characteristic frequencies

If the sample in infrared spectroscopy is in the form of a liquid, then the molecules exchange energy by collision with their neighbours during the period of a single rotation, and rotational fine structure is no longer observed. Instead, a series of bands is observed, a typical spectrum being shown in Figure 6.12. It might be though that the absence of detail would

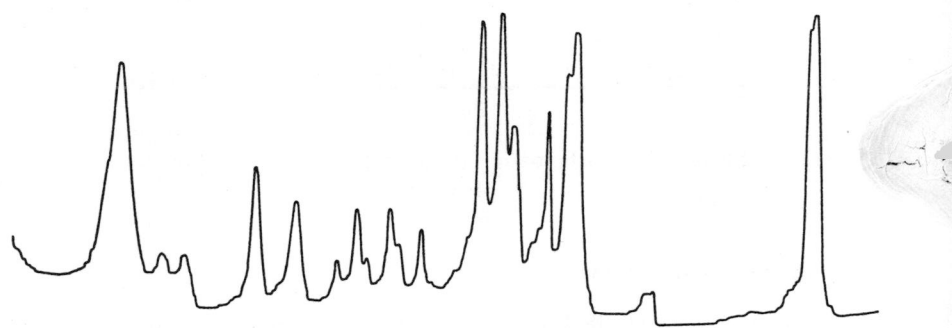

Figure 6.12 A typical ir spectrum

make this spectrum of scant use to the chemist, but in fact nothing could be further from the truth. The chief value of the infrared spectra of complicated molecules arises from the fact that the vibrational frequencies of characteristic groups are relatively insensitive to the nature of the rest of the molecule. Thus the >C=O group in $(CH_3)_2C=O$ has much the same vibrational frequency as that in $(C_2H_5)_2C=O$. Table 6.1 gives some characteristic frequencies. These

Table 6.1   Some characteristic vibrations in infrared spectra

Molecular fragment	Frequency (cm^{-1})
O–H (non-H-bonded)	3650–3600
N–H	3500–3200
=C–H	3100–3000
C–H	2970–2850
–C≡N	2275–2200
–C≡C–H	2260–2100
C=C	1680–1620
C=O	1780–1660
C–F	1400–1000
C–C	1250–700
C–Cl	800–600

frequencies may be used to analyse organic substances for the presence of these groups.

The skeletal vibrations of organic molecules are at rather lower energies in the region 700 to 1400 cm^{-1}. They often produce a complex vibrational spectrum, but even the very complexity of this spectrum can be useful, acting as a 'fingerprint' for the molecule. This is the basis for the widespread use of infrared spectroscopy in identifying molecules.

K.   Experimental infrared spectroscopy

Infrared spectroscopic experiments are almost always absorption experiments, so the first requirement is a source of continuous infrared radiation. Suitable sources are the Nernst filament, which is a spindle of rare earth oxides, or the Globar, which is a rod of carborundum. Both are heated electrically.

The sample must be held in a cell which has windows which are transparent to infrared radiation. Rock salt, NaCl, and potassium bromide are commonly used, and are suitable for most of the infrared region. Their main drawback is that they cannot be used with aqueous solutions. The spectrum may be dispersed by a diffraction grating. Detectors use either the heating effect of the radiation or photoconductivity effects in semi-conductors. Heat may be detected by a Golay cell, where a small temperature rise produces a pressure increase and hence a change in inductance. Alternatively, a bolometer or thermocouple may be used.

A typical experimental arrangement is shown in Figure 6.13, where it can be seen that a differential technique is used. The incident radiation is chopped and sent through two paths, one of which contains the sample, and the resulting difference amplified to work a pen-recorder.

Instruments of this kind which measure accurately the relative intensities of spectral lines are called spectrometers. Instruments which are more sensitive to line position than intensity are called spectographs.

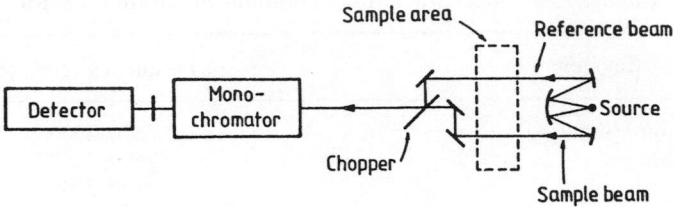

Figure 6.13 Experimental arrangement for infrared spectroscopy

L. The ammonia inversion spectrum

There is one type of vibrational spectrum which we have not yet covered, and
which indeed is not even found in the infrared region of the spectrum. This is
the inversion spectrum, the most studied example of which is ammonia, whose
inversion spectrum is found in the microwave region, at about 0.8 cm^{-1}.

If a graph is plotted of the potential energy of ammonia against the angle θ
shown in Figure 6.14, then a double minimum is obtained, as shown in Figure
6.15. Two vibrational energy levels are shown. The height of the energy barrier
is 2070 cm^{-1}; classically this is the energy required to turn the molecule 'inside
out' rather like an umbrella.

Figure 6.14 Geometry of the NH_3 molecule

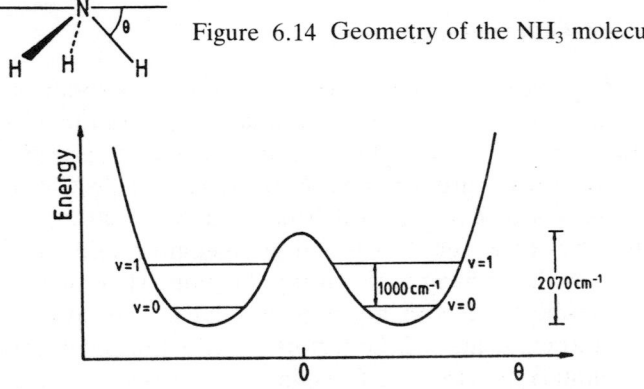

Figure 6.15 Vibrational levels in NH_3 (without
tunnelling)

If the vibrational wavefunctions are drawn in, it is found that because of the
finite height of the barrier, the wavefunctions of the left- and right-hand sides of
the diagram can overlap slightly. This is another example of quantum
mechanical tunnelling. We must now correctly consider linear combinations of
the vibrational wavefunctions, just as we took linear combinations when we
brought together two hydrogen atoms to form H_2. These combinations will
have slightly different energies, as is shown in Figure 6.16. The ammonia

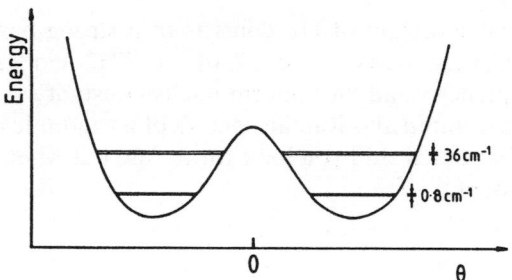

Figure 6.16 Vibrational levels in NH_3 (including tunnelling)

inversion spectrum thus arises from transitions between these two levels, which are split by $0.8\ cm^{-1}$.

M. Summary

Absorption of infrared light produces changes in the vibrational and rotational energies of molecules. In most cases only the lowest vibrational energy state is occupied, the main feature of the spectrum being a band corresponding to the $v = 1 \rightarrow v = 0$ transition. The position of the band gives an approximate value of the vibration frequency, and its rotational fine structure gives accurate information on the bond lengths in the $v = 0$ and $v = 1$ states. Weaker bands due to transitions from $v = 0$ to higher states (overtones) give information on the detailed shape of the potential curve.

Polyatomic molecules have more than one vibrational mode; they are commonly studied both by infrared absorption and Raman spectroscopy. Vibrations are infrared active if they involve a change in dipole moment and Raman active if they involve a change in polarizability. In molecules with a centre of symmetry, each vibrational mode may be observed in infrared or Raman, but not both. Many polyatomic molecules have been studied in solution, where all rotational structure is lost; however, many vibration frequencies are characteristic of particular bonds or groups, and infrared spectra are widely used in analytical work.

N. Problems

1. The $^{16}O_2$ molecule has a vibration frequency of $1580\ cm^{-1}$. Calculate the force constant of the bond. ($^{16}O = 15.995$).
2. In the infrared spectrum of $H^{35}Cl$ the lines in the P branch are observed at 1752.0, 2775.8, 2799.0, 2821.6 cm^{-1}, with R lines at 2906.2, 2925.6, 2945.0 and 2963.3 cm^{-1}. What is the force constant for the bond and the bond lengths in vibrational levels 0 and 1?
3. Which normal modes of the water molecule are active in the infrared and in Raman spectroscopy?

4. The vibrational spectrum of HF consists of a strong band (1–0) at 3958 cm^{-1}, and a very weak band (2–0) at 7737 cm^{-1}. Calculate the vibration frequency ν and the anharmonicity constant x.

5. If you had the infrared and Raman spectra of a triatomic molecule such as HCN, what features would you look for to find out whether the molecule was linear or bent?

Chapter 7
Electronic Spectra

Transitions between electronic states of molecules generally occur in the visible and ultraviolet regions of the spectrum. Whereas rotation and vibration spectra are almost always studied in absorption, electronic spectra are commonly studied in both absorption and emission, and we shall see that these two techniques are often complementary.

When a molecule changes its electronic state, it may also change its vibrational and rotational states, and an electronic transition will therefore have superimposed on it both vibrational and rotational fine structure. As the rotational structure is only observable in very small molecules, diatomic molecules have been most extensively studied.

Unlike the vibrational and rotational energy levels, there are no simple relationships between the electronic energy levels of molecules, and one of the main objects of electronic spectroscopy is to provide information on the energies of the various states.

A. Absorption spectra of diatomic molecules

Figure 7.1 shows a typical potential energy diagram for an electronic transition in a diatomic molecule. Here the vibrational energy levels are shown, but not the rotational levels. If the experiment is carried out at room temperature, then only the $v = 0$ level of the lower state will be appreciably populated (just as in vibrational spectroscopy). Now in electronic spectra, there is no simple selection rule governing the change in vibrational quantum number, so transitions to all the vibrational levels of the upper state may be observed in principle. Thus at low resolution the absorption spectrum will consist of a series of peaks, corresponding to the vibrational transitions $0 \rightarrow 0$, $0 \rightarrow 1$, $0 \rightarrow 2$, etc.

The differences in vibrational energy in the upper state between the levels with quantum numbers v and 0 is

$$\Delta E_{vib} = h\nu_{vib}(v + \tfrac{1}{2}) - xh\nu_{vib}(v + \tfrac{1}{2})^2 - \tfrac{1}{2}h\nu_{vib} + \tfrac{1}{4}xh\nu_{vib}$$
$$= h\nu_{vib}v - hx\nu_{vib}v(v + 1).$$

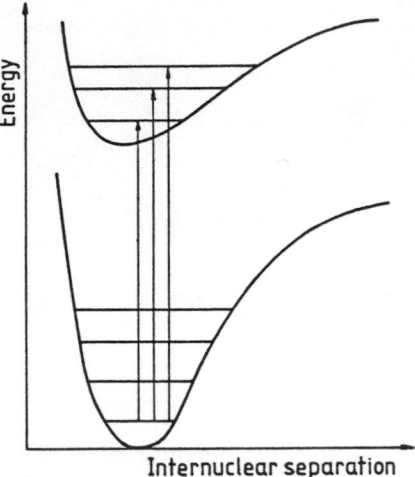

Figure 7.1 An electronic transition of a
diatomic molecule

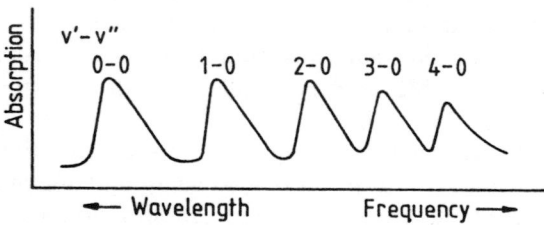

Figure 7.2 A low re-
solution   electronic
spectrum

Therefore the frequencies of the peaks in the electronic spectrum are given by

$$\nu = \nu_{00} + \nu_{vib} - \nu_{vib}x\upsilon(\upsilon + 1),$$

where ν_{00} is the frequency of the $0 \to 0$ transition. This is shown schematically in Figure 7.2. Thus the absorption spectrum gives us information on ν_{00}, the excitation energy of the upper state, and also on ν_{vib} and x, the vibrational constants of the upper state.

B.   Rotational fine structure

If each of the peaks in the electronic spectrum is studied at higher resolution, it is found to consist of a series of lines, which is due to changes in the rotational quantum number. The structure is very similar to that seen in infrared spectroscopy. The energy levels for one vibrational transition are shown in Figure 7.3. Thus the formulae which we have previously derived for P, Q and R branches in Chapter 6 still apply, with ν_{vib} being replaced by $(\nu_{00} + \Delta\nu_{vib})$, the change in electronic and vibrational energy. P and R branches ($\Delta J = -1$ and

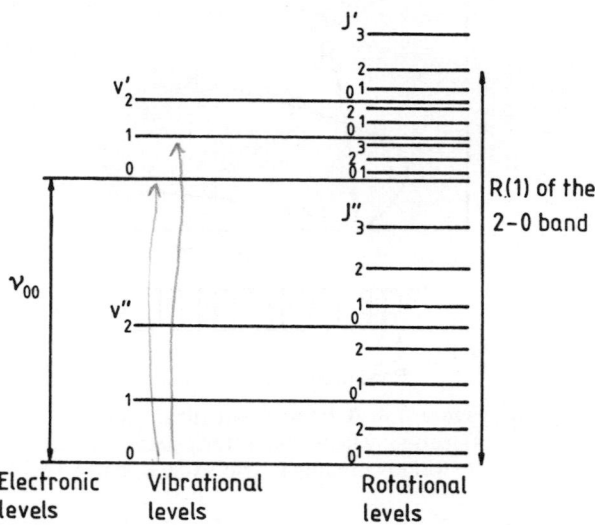

Figure 7.3 Energy levels involved in the rotational structure of an electronic transition

+1) are observed for all diatomic molecules; Q branches ($\Delta J = 0$) are observed only if there is a change in electronic orbital angular momentum between the two states (e.g. in a $^1\Pi \rightarrow {}^1\Sigma$ transition).

In Chapter 6 we saw that the equation giving the frequencies of lines in the R branch is

$$h\nu = \Delta E_{el} + \Delta E_{vib} + 2B' + \underset{-ve}{(B' - B'')J^2} + \underset{+ve}{(3B - B'')J}$$

Thus for low values of J the lines are roughly equally spaced, but as J increases the J^2 term causes the lines to converge, as in vibrational spectra $B'' > B'$. However, this convergence is slow, and not observed in many cases. In electronic spectra, the difference between B' and B'' may be very much larger, as the internuclear separations in different electronic states can be widely different. If the internuclear separation is greater in the upper state, then $B' < B''$ (remember $B \propto 1/r^2$), and the R branch will show a band-head, as in Figure 7.4. As the difference in B values is large, the convergence is rapid, and the band-head is usually observed.

However, it is also possible that the internuclear separation is less in the upper state than in the lower state. Then $B' > B''$, and clearly the R branch lines do not converge at all. However, for the P branch

$$h\nu = \Delta E_{el} + \Delta E_{vib} + (B' - B'')J^2 - (B' + B'')J,$$

and now the coefficients of J and J^2 have opposite signs, and so the P branch will show a band-head. This is shown in Figure 7.5. Thus the observation of band-heads immediately tells us whether the internuclear separation is larger

82

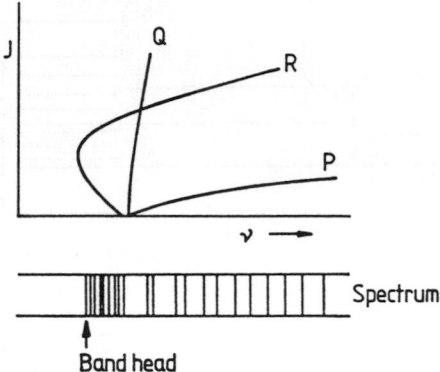

Figure 7.4 A band head in an electronic transition where the internuclear separation is greater in the excited state

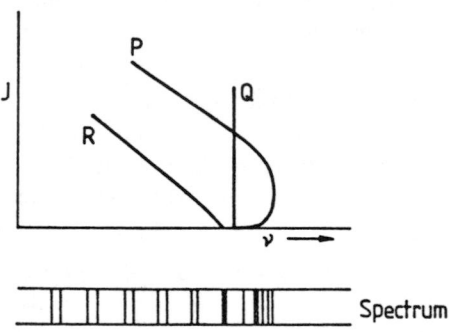

Figure 7.5 The band head where the internuclear separation is less in the excited state

in the ground state or the upper state, and this in turn suggests whether a bonding electron (producing an increase in bond length) or an antibonding electron (producing a decrease in bond length) is being excited.

A band such as that shown in Figure 7.4 shows a definite cut-off at high frequencies and a tail to low frequencies, and is said to be red-degraded. One such as that in Figure 7.5, where the tail is now towards high frequencies (the blue end of the spectrum), is said to be blue-degraded or violet-degraded.

An absorption spectrum at room temperature thus gives us information on the rotational constants of the $v = 0$ level of the ground electronic state and on the rotational constants of many of the vibrational levels of the excited electronic state. Furthermore, the presence of a Q branch gives us information on the nature of the electronic transition involved.

C.   Emission spectra

Electronic spectra are frequently observed in emission rather than absorption; as electronic excitation energies are typically much larger than thermal energies, at room temperature, it is necessary to provide energy to raise molecules to their excited states. This is commonly achieved by electric discharges or by simple heating.

The general appearance of an emission spectrum is more complex than that of the corresponding absorption spectrum, even if only one electronic transition is considered. In the absorption experiment, normally only the lowest vibrational energy level is occupied, whereas in emission experiments it is common for many vibrational levels to be occupied. This difference is illustrated in Figure 7.6.

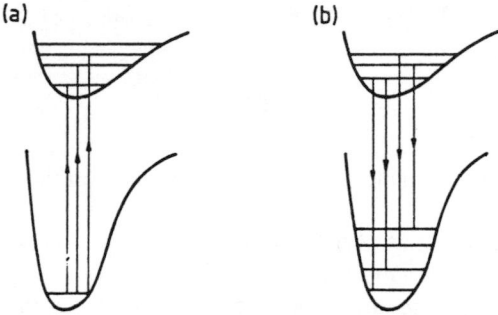

Figure 7.6 Absorption and emission, showing the vibrational levels involved

There is no general selection rule governing the change in vibrational quantum number, and transitions between many values of v'' and v' may be observed. Each of these transitions gives rise to a band, which of course shows rotational fine structure as described in the previous section. The intensities of these bands are governed by the Franck–Condon principle, which is described in Chapter 8.

The frequencies or intensities of the observed bands are often shown in a Deslandres table, which gives the quantum number of the upper and lower vibrational levels. A Deslandres table for CO is given in Figure 7.7. This type of table may also be used to give the intensities of various bands.

Although the emission spectrum may be rather more complex than the absorption spectrum, as transitions are observed to many vibrational levels of the lower state, it is possible to obtain both vibrational and rotational constants for the vibrational levels of the ground state. A further advantage of the emission spectrum is that it may be possible to observe transitions between two excited electronic states. In absorption spectra, one of the electronic states must be the ground state, but this is not the case in emission spectra, which may prove to be the only source of information on some excited states.

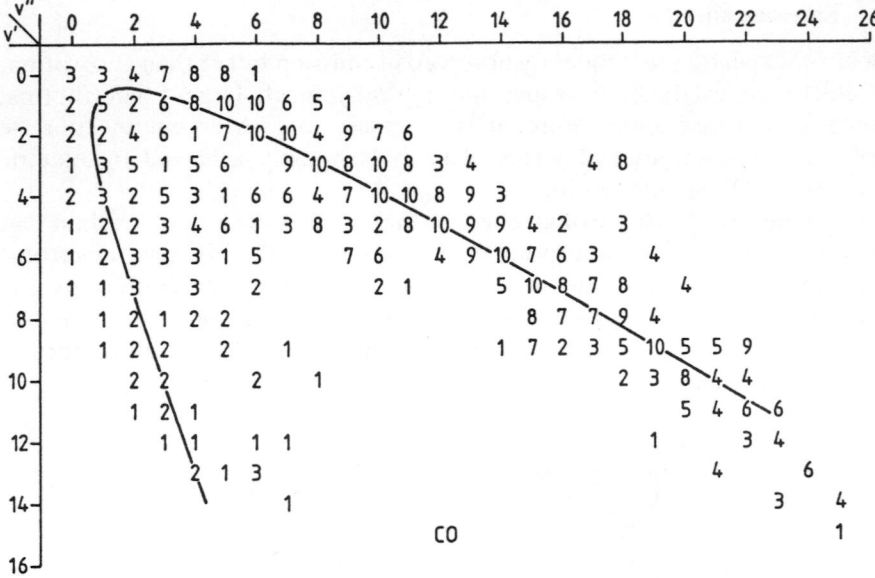

Figure 7.7 A Deslandres table showing bands observed in the spectrum of CO

In a sense emission and absorption electronic spectra provide complementary information: the former reveals information about the potential minimum of the upper state and the latter about the ground electronic state.

D. Vibrational and rotational numbering

In a favourable case, there may be no difficulty in identifying the vibrational quantum numbers of each band in an electronic spectrum. However, in other cases the intensity of a series of bands may drop to zero before the $v = 0$ band is reached, and here the relative numbering may be established, but not the absolute numbering.

The absolute numbering of the vibrational levels may be achieved by using the isotope effect. Two molecules which are isotopically related have identical potential curves, but their vibrational energy levels differ as they have different reduced masses. We recall that for an harmonic oscillator:

$$\nu_{vib} = \frac{1}{2\pi} \sqrt{\frac{k}{\mu}},$$

and so for two isotopic species I and II,

$$\frac{\nu_I}{\nu_{II}} = \sqrt{\frac{\mu_{II}}{\mu_I}} = \rho,$$

where $\rho^2 = \mu_{II}/\mu_I$. The energy levels of the heavier molecules show the same pattern as the lighter molecule, but are less widely spaced. This means that the heavier molecule has its bands shifted relative to the higher molecule, and this shift varies with v. Thus if the shift is calculated for the various values of v, it may be compared with the experimentally determined shifts, allowing definite identification of the vibrational quantum number.

Spectroscopists must also try to assign J values to all the lines in the rotational fine structure of an electronic spectrum. Once again, it is often easy to identify a series of lines whose relative numbering is obvious, but the band origin may not be clear and the absolute J values more difficult to determine. The method of combination differences which are encountered in infrared spectra can be useful here; the combination difference formulae were

$$\Delta_2 F'' = R(J - 1) - P(J + 1) = 4B''(J + \tfrac{1}{2}) - 8D''(J + \tfrac{1}{2})^3$$

and

$$\Delta_2 F' = R(J) - P(J) = 4B'(J + \tfrac{1}{2}) - 8D'(J + \tfrac{1}{2})^3.$$

In this method, $\Delta_2 F/(J + \tfrac{1}{2})$ is plotted against $(J + \tfrac{1}{2})^2$; if the rotational numbering is correct, then a straight line is produced, but if the numbering is slightly wrong, then curves are produced as in Figure 7.8. A further check may be obtained by determining the value of D from the gradient of the graph; it should be close to the value of D given by the Kratzer relationship which we met in Chapter 5:

$$D = \frac{4B^3}{\omega^2}.$$

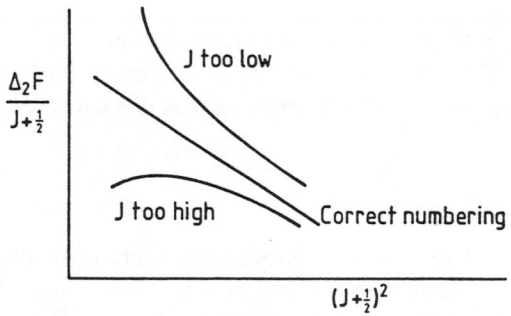

Figure 7.8 Determination of rotational numbering

Further confirmation of the rotational numbering may be obtained from the study of isotopic species; it may be shown that the rotational constants of the isotopic species are related to ρ, the ratio of the reduced masses, by the formulae

$$B_{II} = B_I \rho^2 \quad \text{and} \quad D_{II} = D_I \rho^4.$$

E. Electronic states

Electronic spectroscopy of diatomic molecules has two main aims. One is the provision of structural data, giving a detailed understanding of vibrational and rotational constants. The second aim is to provide information on the electronic structure of the molecules, to allow us to answer questions about why molecules form and how they are held together. Here molecular spectroscopy gives us information about molecular orbitals just as atomic spectroscopy does about atomic orbitals.

In Chapter 2 we discussed molecular orbital diagrams and the various electronic states which could arise from particular electron configurations. Detailed study of rotational fine structure allows us to deduce the symmetries of the electronic state involved in a transition; the chemist must then assign the observed states to molecular orbital configurations. One of the problems in assigning the states is that many electronic states may not be seen in the spectrum; there are two important selection rules which hold well for light molecules, though less well for some heavier molecules. They are

$$\Delta\Sigma = 0 \quad \text{and} \quad \Delta\Lambda = 0, \pm 1.$$

The first rule means that the total electron spin never changes in an electronic transition; if a molecule has a singlet ground state, i.e. with all its electrons paired, then the absorption spectrum will only show transitions to other singlet states. Triplet states, in which two electrons have the same spin, will only be seen in the emission spectrum, and even then they will not give rise to transitions to the ground state. The second rule means that transitions such as $^1\Sigma \rightarrow {}^1\Sigma$, $^2\Sigma \rightarrow {}^2\Pi$ and $^3\Pi \rightarrow {}^3\Delta$ are known, but not $^1\Sigma \rightarrow {}^1\Delta$ (except with very low intensity).

Many diatomic molecules have their states definitely assigned to configurations. However, there are still some cases of poorly understood molecules, either where there is little data or where the states cannot be unambiguously assigned.

F. Dissociation Energies

The most accurate values of the dissociation energies of diatomic molecules come from their electronic spectra. It is important to realize that there are two different definitions of dissociation energy, as shown in Figure 7.9. They differ only by the zero-point vibrational energy, which may be found from the vibrational constants. There are two important spectroscopic methods of finding dissociation constants: the observation of a continuum limit and Birge–Sponer extrapolations.

Observation of a continuum limit

In a favourable case, such as I_2, it may be possible to observe vibrational structure right up to the onset of dissociation (Figure 7.10). The vibrational

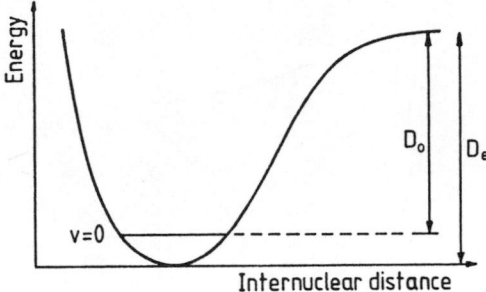

Figure 7.9 Dissociation energies D_o and D_e

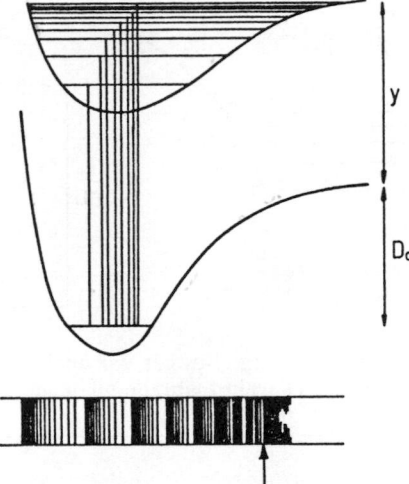

Figure 7.10 Observation of the onset of
a dissociation continuum

bands converge as the dissociation limit is approached; if the incident photon energy is greater than the dissociation energy D_0 then the atoms separate, and as translational energy is not quantized there is continuous absorption of energy in the spectrum beyond the dissociation limit.

The energy required to excite a molecule from its ground state up to the dissociation limit of its upper state is clearly equal to $(y + D_0)$, where y is the energy separation of the appropriate atomic energy levels. Thus D_0 can be determined if y is known. The value of y is readily obtained from atomic spectra, provided that the relevant atomic states are known. This problem may be overcome if an approximate value of D_0 is known, perhaps from mass spectrometry. Alternatively, the Wigner–Witmer correlation rules may be used. These rules predict what molecular states can arise from combinations of atoms in specified states. Essentially they are conservation of angular

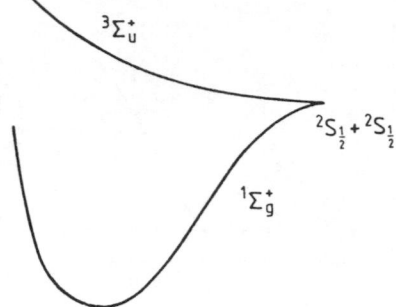

Figure 7.11 Molecular states arising from ^2S atoms

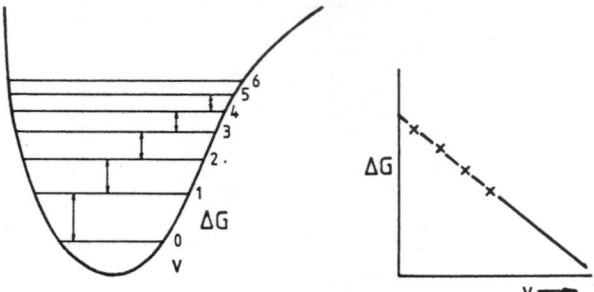

Figure 7.12 The Birge–Sponer extrapolation of vibrational spacing ΔG against vibrational quantum number v

momentum rules. If two atoms in ^1S states (i.e. with no spin or orbital angular momentum) combine, then as they have no orbital angular momentum, they must give a Σ state, but it could be $^1\Sigma$ or $^3\Sigma$. Hydrogen is a simple example of this, and is shown in Figure 7.11.

Birge–Sponer extrapolations

If vibrational levels are not observed right up to the dissociation limit, it may still be possible to extrapolate the vibrational energy levels which are observed. This is shown in Figure 7.12. The limiting value of v may be fitted into the equation

$$D_0 = h\nu_{vib}(v + \tfrac{1}{2}) - h\nu_{vib}(v + \tfrac{1}{2})^2.$$

This extrapolation is really only useful if a significant number of vibrational levels are known; long Birge–Sponer extrapolations are unreliable, and the higher vibrational levels may not continue trends observed for the lower levels.

G. Predissociation

In some instances continuous spectra are observed before excitation to a dissociation limit. This may be caused by predissociation. When a molecule is excited to a point in the excited electronic state which is above the crossing point of the unbound curve, then the molecule may, if selection rules permit, cross to the unbound curve giving a continuous spectrum. We may thus observe perfectly normal spectral lines indicating a progression towards dissociation, but suddenly the fine structure disappears. In absorption the levels just above the crossing point may still give lines, but these are broadened due to the lifetime of the levels being shortened. In emission, since there is no population of levels above the crossing point, the lines involving transitions with these levels simply do not appear and there will be a sharp cut-off in fine structure.

In some cases crossing to a particular unbound curve which crosses a bound state may not be allowed, perhaps for reasons of symmetry. Predissociation should not then occur, but it may often be induced by some external influence, perhaps by collisions which may alter the symmetry of the situation or by external fields.

H. Resonance fluorescence

When a molecule absorbs energy by absorbing a photon, it may eventually lose that energy either by non-radiative processes, e.g. losing its energy to colliding molecules, or by re-radiating the energy. If the energy is re-radiated, the phenomenon is known as fluorescence. In condensed phases the excited molecule normally loses its vibrational energy rapidly to other molecules, and then radiates energy by transitions to various vibrational levels of the ground state.

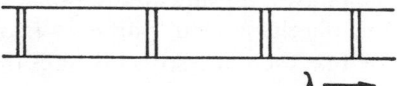

Figure 7.13 A resonance fluorescence
spectrum

If the molecule is in the gas phase at low pressure, it is unable to lose its energy to other molecules, if the radiation rate is greater than the collision rate. Furthermore, if the exciting light is sharply monochromatic, e.g. from a laser, then just one level with a particular value of v and J will be produced. This level can re-radiate light, subject to the selection rule $\Delta J = \pm 1$. There is no restriction on the change in v, and so the resulting spectrum will consist of pairs of lines, each pair corresponding to a particular vibrational level. This simple spectrum may be analysed to give spectroscopic constants for the lower state; such an experiment is called resonance fluoresence. Figure 7.13 shows a typical spectrum. The splitting of each of the pairs of lines is the difference between two rotational levels ($J + 1$ and $J - 1$), and the spacing between the pairs corresponds to differences in vibrational energy levels.

I. Experimental techniques

For absorption spectroscopy the molecules must be held in a tube, with windows which are transparent to ultraviolet light. Glass or quartz may conveniently be used; if the molecules are rather involatile, the tube may be heated. The best source of continuous radiation is a high pressure xenon lamp; the ionization continuum of the xenon atom stretches right into the visible region of the spectrum, and the high pressure broadens the few atomic lines which could confuse the spectrum. The light may be dispersed using a diffraction grating or a prism, and detected either photographically or using a photomultiplier.

In emission spectroscopy it is necessary to excite the molecules so that they can emit light. The necessary energy may be provided by an electric discharge, although this tends to produce 'dirty' spectra, with bands due to impurities. Other methods include microwaves or radiofrequency sources; the quanta provided by these can only increase the translational and rotational energies of the molecules, but if sufficient power is used, collisions will produce electronic excitation. Electrical heating, flames and hollow cathodes have been used to excite molecules, and in fluorescence light itself is used.

J. Electronic spectra of polyatomic molecules

The electronic spectra of polyatomic molecules are very rich in information but extraction of this information may be very complex. Many polyatomic molecules have spectra which do not exhibit all the fine structural detail commonly observed with diatomic molecules. This may be caused by a number of factors. Molecules in condensed phases showing no rotational structure are considered later in Chapter 10. Even in the gaseous phase we may still not see rotational lines. One cause can be that there are so many lines that their linewidths may exceed the separation from their neighbours, giving an unresolvable spectrum. Again, in polyatomic molecules it is often the case that part of the molecule may be dissociated, giving a continuous rather than a discrete spectrum.

These restrictions mean that relatively few molecules have had their spectra resolved and analysed to the level of assigning all the rotational transitions. Where spectra are seen they follow the diatomic pattern: an electronic transition with vibrational and rotational changes superimposed. However, since there may be several vibrational modes and more than one rotational quantum number may change, the resulting spectra are often bewilderingly complex.

Most electronic spectra of polyatomic species have been studied in absorption, but a few have been obtained in emission in electric discharges, in fluorescence and in flames. A particularly fruitful source of absorption spectra of polyatomic free radicals has been the flash photolysis technique illustrated in Figure 7.14. In essence, molecules are dissociated into radicals by a pulse of light

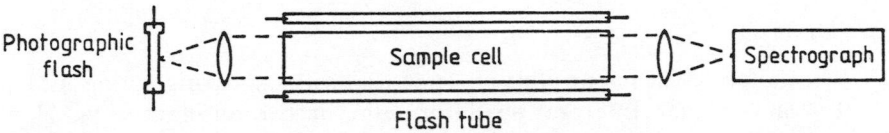

Figure 7.14 Schematic diagram of flash photolysis with the photographic flash being initiated following a delay after the flash tubes

and then after a suitable short delay of the order of milliseconds a second flash of continuous light acts as the source for an absorption spectrum.

Generally, apart from the free radicals the information about ground states of polyatomic molecules is more easily and more accurately obtained from infrared, Raman or microwave spectra so that electronic spectra are studied largely for what they reveal about the geometrical structure of excited electronic states and about their ionization potentials and dissociation energies.

K. Rydberg transitions

Both diatomic molecules and polyatomic molecules may have electronic transitions where the excited electron is removed to outer orbitals which are so far from the molecule that the electron will only experience an electric field acting as if it were crudely at a central point. In such circumstances the situation would be very reminiscent of the excitation of electrons in atoms. The orbitals concerned are very diffuse and have energies which will follow the atomic pattern, approximately fitting a formula $E \approx R/n^2$, with R being the Rydberg constant. The electronic levels are thus called Rydberg levels.

Since the levels may be fitted to the simple formula appropriate for atomic energy levels, it is possible to use such a formula to extrapolate the energy levels to the point where the excited electron is completely removed from the molecular framework. Until the advent of photoelectron spectroscopy this was a major source of information on molecular ionization potentials.

L. Summary

Electronic spectra arise from moving an electron from one molecular orbital to another. The relative energies of molecular orbitals are obvious products from studying the spectra, but in addition the changes in vibrational and rotational energy yield information of vibration frequencies, dissociation energies and bond lengths. These data are available for the electronic ground state and for excited states, as spectra may be observed in absorption and emission. Although richer than microwave or infrared spectra, electronic spectra can only be resolved and analysed fully for quite small molecules.

92

M. Problems

1. The molecule CO has an emission spectrum with bands originating at 115, 154 and 451 nm. Interpret this information in two different ways. How would the absorption spectrum distinguish between your two suggestions?

2. The following are observed wave numbers of lines in the 0–0 band of BeO

J	$R(J)$	$P(J)$
0	21 199.8	
1	202.9	21 193.3
2	205.7	189.9
3	208.5	186.4
4	211.1	182.7
5	213.6	178.9
6	215.6	174.8
7		170.7

How can you confirm the assignment of the rotational numbering? What are the B values for the two vibrational levels?

3. Describe the appearance of the electronic emission spectrum involving a stable upper electronic state of a diatomic molecule and a lower state which is (a) unstable at all internuclear separations (b) stable but predissociated at large internuclear distances.

4. Given the following spectroscopic data for ground state iodine (in cm^{-1}), $\omega_e = 214.6$, $x_e\omega_e = 0.6$, $B_e = 0.0374$, $\alpha_e = 0.0001$, calculate the resonance fluorescence spectrum observed when iodine is excited by a laser to a single rotational level $v' = 2, J' = 10$.

5. (a) The molecule Br_2 has a dissociation energy $D_o = 1.971$ eV. Its vibration frequency is 323 cm^{-1}. Calculate the value of the dissociation energy D_e in cm^{-1} (1 eV = 8065 cm^{-1}).
   (b) A series of absorption bands is observed in the ultraviolet spectrum of O_2. The origins of the first three bands are at 49,363, 50,046 and 50,710 cm^{-1}. Sketch an energy level diagram for the transitions (all of which originate from the $v = 0$ level of the ground state) and estimate the dissociation D_o of the excited state. The dissociation energy is actually 7194 cm^{-1}; comment.

Chapter 8
Intensities

A. Absorption of radiation

Solution of the Schrödinger equation gives information on the allowed rotational, vibrational and electronic states of molecules, and on the energy differences between these states. This alone, however, is not sufficient to explain the appearance of a spectrum, because not all the possible transitions between states are allowed and the transitions which are allowed are not all likely to occur. The rate of absorption of radiation clearly depends on the number of molecules in the appropriate initial state, the energy density of radiation of the appropriate frequency and the intrinsic probability of the transition. Therefore:

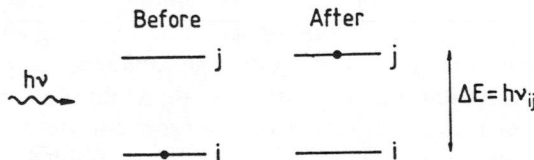

Figure 8.1 The mechanism of absorption

Rate of absorption $= \rho(\nu_{ij}) \, B_{ij} \, n_i$.

Here B_{ij} is called the Einstein B coefficient, and measures the intrinsic probability of the transition.

The problem of finding the value of the coefficient B_{ij} for a particular transition may be approached either experimentally or theoretically. If light is passed through a sample of gas which can undergo only one transition and the amount of light transmitted is measured as a function of frequency, then a spectrum such as that in Figure 8.2 is obtained. The fraction of light absorbed is not a very useful measure of the probability of the transition, as it depends on the gas concentration, c, and the distance that the light travels through the gas,

94

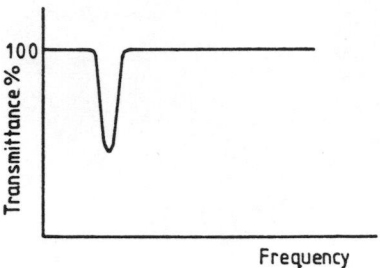

Figure 8.2 The absorption of radiation

l. The Beer–Lambert law describes the variation in intensity of the transmitted light, I, with c and l:

$$\log_{10}I = \log_{10}I_0 - \varepsilon cl,$$

where I_0 is the intensity of the light in the absence of the sample and ε is now a constant for the particular transition and is called the extinction coefficient. When c is measured in moles per litre and l in centimetres, ε is called the molar absorption coefficient; values vary from 10^5 or more for the strongest absorptions to 1 or less for the very weak absorptions. It may be shown that ε is related to B_{ij}:

$$\varepsilon \propto \nu_{ij}\, B_{ij}$$

Alternatively, values of B_{ij} may be derived from the wavefunctions of the initial state i and the final state j. Light is a form of electromagnetic radiation with mutually perpendicular oscillating electric and magnetic fields. The interaction of the electric field with atoms and molecules is generally of the order of one hundred thousand times stronger than that of the magnetic field, and it is on this electric interaction that we shall concentrate.

Classically the energy of interaction of an electric field E with an electric dipole of moment μ is given by:

$$\text{Energy} = E\,\mu.$$

In quantum mechanics the dipole moment of a molecule is calculated by evaluating the integral:

$$\mu \int \Psi^*(\textstyle\sum_n e_n r_n)\Psi \; d\tau,$$

where e_n is the charge on the nth particle and r_n is a vector representing its position. This may be written

$$\mu = \int \Psi^* R \Psi \; d\tau,$$

where R is called the dipole moment operator and is a vector quantity.

Einstein showed that the relationship between B_{ij} and the initial wavefunctions is

$$B_{ij} = c\left|\int\Psi_i^* R\Psi_j\, d\tau\right|^2$$

where c is a constant. The integral $\int\Psi_i R\Psi_j\, d\tau$ is often referred to as a matrix element of the dipole moment operator, and may be written in shorthand as R_{ij}. The Dirac notation is also commonly used:

$$R_{ij} = \int\Psi_i^* R\Psi_j\, d\tau = \langle\Psi_i|R|\Psi_j\rangle.$$

In principle it is therefore possible to calculate the rate of absorption of radiation by a particular transition if the wavefunctions of the initial and final states and the concentration of molecules in the initial state are known.

B. Population effects

Small molecules may exist in a variety of different states, each of which may be described by rotational, vibrational and electronic quantum numbers. If a collection of molecules is at thermal equilibrium, then the number of molecules in each state is given by the Boltzmann distribution law:

$$N_i = N_0\, e^{-\varepsilon_i/kT}.$$

where N_i is the number of molecules in state i, ε_i is the energy of state i relative to the lowest state and N_0 is the number of particles in the lowest state. k is the Boltzmann constant, which has the value 1.38×10^{-13} J mol^{-1}K^{-1}, and T is the temperature in kelvin. Here it is important to count each state separately; if a state has degeneracy p_i, then the Boltzmann law becomes

$$N_i = p_i\, N_0\, e^{-\varepsilon_i/kT}.$$

We can consider the application of the Boltzmann law to rotational, vibrational and electronic energy levels in turn.

Rotational levels

We have seen that the rotational levels of a linear molecule are described by the rotational quantum number J and that the energy of each level is given by

$$E_J = BJ(J + 1).$$

In addition to the quantum number J, its component M_J is also defined, although in the absence of an external field, states with different values of M_J are all degenerate. M_J may take the values of $J, J - 1,..., -(J - 1), -J$; thus the total number of values of M_J is $(2J + 1)$ and each rotational level has a degeneracy of $(2J + 1)$.

The Boltzmann distribution law is now

$$N_J = N_0\, (2J + 1)\, exp\left[\frac{-BJ(J + 1)}{kT}\right].$$

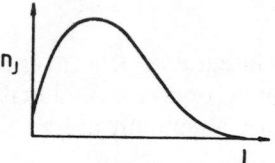

Figure 8.3 The popula-
tion of rotational levels

As the value of B is typically smaller than kT, many different rotational levels are occupied; this is illustrated in Figure 8.3. Note that the peak of this curve does not occur at $J = 0$, even though this is the state of lowest energy, because of the greater degeneracy of higher states. By differentiation it may be shown that the quantum number of the state with the greatest population is

$$J_{max} = \sqrt{\left(\frac{kT}{2B}\right)} - \tfrac{1}{2}.$$

For the hydrogen halides, which have a very high value of B, J_{max} may have a value of only 2 at room temperature, whereas for a diatomic molecule such as CO, $J_{max} = 7$, and for a triatomic molecule such as OCS, $J_{max} = 22$.

Vibrational levels

The energies of allowed vibrational levels are, taking account of anharmonicity, given by

$$E = (v + \tfrac{1}{2})\omega_e - \omega_e x_e(v + \tfrac{1}{2})^2.$$

Therefore the difference in energy between the states $v = 0$ and $v = 1$ is

$$\varepsilon_1 = \omega_e - 2\omega_e x_e$$

and the population of the state $v = 1$ is given by

$$N_1 = N_0 \exp\left[\frac{-(\omega_e - 2\omega_e x_e)}{kT}\right].$$

For a typical example such as HCl, $\varepsilon_1 = 2900$ cm^{-1}, and so at room temperature

$$\frac{N_1}{N_0} = \exp\frac{-(2900 \times 3 \times 10^{10} \times 6.63 \times 10^{-34})}{1.38 \times 10^{-23} \times 300}$$
$$= 9 \times 10^{-7}$$

Clearly in this case the population of all vibrational levels other than the lowest is negligible, and so absorption from $v = 0$ only is observed. Similarly, in rotational–vibrational Raman spectroscopy, anti-Stokes bands are much weaker than the Stokes bands. Table 8.1 shows the variation in population of the $v = 1$ state with ω_e and temperature. The population of the $v = 1$ level

Table 8.1 Values of N_1/N_0

$\omega_e(cm^{-1})$	Temperature (K)		
	300	600	1200
4000	5×10^{-9}	7×10^{-5}	8×10^{-3}
3000	6×10^{-7}	7×10^{-4}	3×10^{-2}
2000	7×10^{-5}	8×10^{-3}	9×10^{-2}
1000	8×10^{-3}	9×10^{-2}	3×10^{-1}

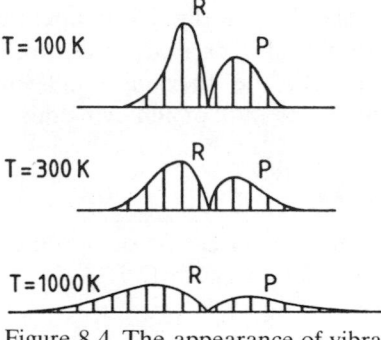

Figure 8.4 The appearance of vibration–rotation bands as a function of
temperature

becomes important when the bond is weak or the temperature high; the observed spectrum will then contain a band of low intensity due to the $v = 1$ to $v = 2$ transition. This transition occurs at

$$\Delta E = E_2 - E_1$$

$$= (2 + \tfrac{1}{2})\omega_e - (2 + \tfrac{1}{2})^2\omega_e x_e - (1 + \tfrac{1}{2})\omega_e - (1 + \tfrac{1}{2})^2\omega_e x_e$$

$$= \omega_e - 4\omega_e x_e,$$

i.e. at a slightly lower energy than the main $v = 0 \rightarrow v = 1$ band. This band is called a hot band, and may be recognized easily as its intensity varies rapidly with temperature.

If we combine the rotational populations with the vibrational, then we can describe the expected appearance of rotation–vibration bands. Figure 8.4 shows schematically the variation of R and P branches as a function of temperature. This temperature variation can be of value in two ways. If the temperature of a source is known, then the separation of the intensity maxima of the P and R branches can give a value of the rotational constant B, even if individual lines cannot be resolved. It is straightforward to show

$$\Delta\nu_{PR} = \left(\frac{8BkT}{hc}\right)^{1/2}.$$

Conversely, this effect may be used to measure temperature if the rotational constant B is known. This is important in astrophysics and may be used to find the temperatures of planetary atmospheres. Here the rotational structure must be sufficiently well defined for the molecule to be identified, of course.

Electronic levels

There is no simple formula which gives the energies of different electronic states in molecules, but in the great majority of cases the energy difference between the lowest state and all others is so large that the populations of the excited states are negligible at all reasonable temperatures. When significant numbers of molecules are found in excited states, this is normally a sign that some process which disturbs the thermal equilibrium is occurring; the absorption of light is a simple method of achieving this.

C. Franck–Condon factors

We saw in Section A that the intensity of a transition is related to the wavefunctions of the initial and final states by

$$B_{ij} = c \left| \int \Psi_i R \Psi_j d\tau \right|^2.$$

However, it would be quite incorrect to assume that the calculation of intensities is therefore a routine or trivial matter; the evaluation of Ψ_i and Ψ_j is an extremely complex procedure, and accurate wavefunctions are known only for the states of a few very simple molecules.

Fortunately it is possible to make many predictions about the intensities of transitions even when accurate wavefunctions are not available. If we write the total wavefunction as a product of rotational, vibrational and electronic parts,

$$\Psi = \psi_{rot} \times \psi_{vib} \times \psi_{el},$$

then the rotational part ψ_{rot} may be calculated readily if we assume that the molecule is a rigid rotator, and the vibrational part ψ_{vib} may be calculated if we assume that the molecule behaves as an harmonic oscillator, provided that we may take molecular parameters such as bond lengths and force constants as known. The accurate calculation of ψ_{el} is usually difficult, although approximate solutions are often available. The integral R_{ij} may be factorized into rotational, vibrational and electronic parts, and it is then possible to compare the relative intensities of different rotational and vibrational transitions between the same electronic states.

This can be illustrated if we consider the appearance of the electronic absorption spectrum of a diatomic molecule. Labelling the initial and final states i and j, and ignoring the rotation of the molecule,

$$R_{ij} = \int \Psi_i R \Psi_j d\tau$$

$$= \int (\psi_{el})_i (\psi_{vib})_i R (\psi_{el})_j (\psi_{vib})_j d\tau.$$

To a very good approximation, this may be factorized to

$$R_{ij} = \int (\psi_{el})_i R (\psi_{el})_j d\tau \times \int (\psi_{vib})_i (\psi_{vib})_j d\tau.$$

The first term is constant provided that we consider a particular electronic transition, and so the intensities of the various vibrational bands are determined by the term $\int (\psi_{vib})_i (\psi_{vib})_j d\tau$, the vibrational overlap integral. There is no simple formula for this integral as the average bond length and the vibration frequency will be different for the upper and lower electronic states, but it is not difficult to calculate overlap integrals and tabulations exist.

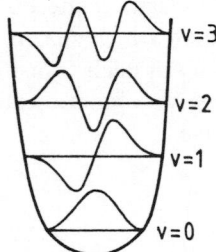

Figure 8.5 Harmonic oscillator wavefunctions

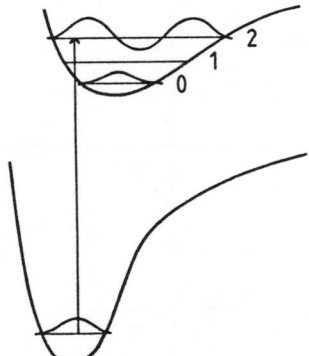

Figure 8.6   Franck–Condon transitions: large overlap for the 2–0 band

We may gain more insight into these vibrational overlap integrals if we consider the problem semi-classically. The Franck–Condon principle states that the time taken for a transition to occur is very short compared to the period of vibration of a molecule, and so the internuclear separation remains constant while the transition occurs. Figure 8.5 shows the vibrational wavefunctions for an harmonic oscillator; note that for low values of v the molecule is most likely to have an internuclear spacing close to r_0, whereas for high values of v an internuclear spacing which is much higher or much lower is more likely. Figure

100

8.6 shows typical potential curves for a diatomic molecule; notice that the bond lengths and force constants are not the same for the two electronic states.

According to the Franck–Condon principle, a transition occurs at constant r and must therefore be represented by a vertical line on this diagram; if we consider absorption from the $v = 0$ state, then the most probable internuclear spacing is close to r_0. The probabilities of transition to the upper vibrational levels $v' = 0,1,2,...$ depend on the magnitude of the upper vibrational wavefunction at the distance r_0. Thus in Figure 8.6 it can be seen that the transition to $v' = 0$ is rather unlikely, and the most probable transition is to the $v' = 2$ level. Transitions to the $v' = 1$ and 3 levels would also be quite likely, but other transitions are less probable. Of course, there is also a very small chance that at the moment of transition the bond length in the ground state might have been much less than r_0; then transition to $v' = 8$ might have occurred. Overall the probability of each transition depends on the degree of overlap between the vibrational wavefunctions of the upper and lower states, as we stated more formally above.

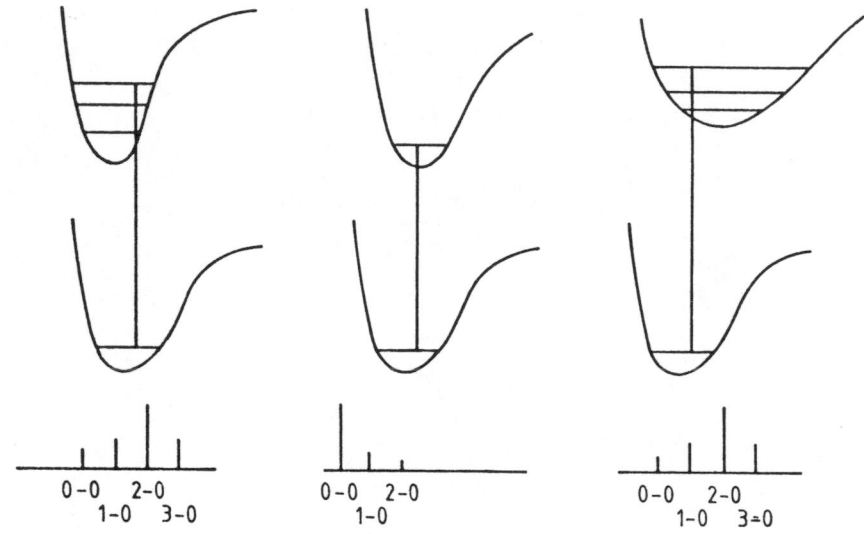

Figure 8.7 The variation in intensity of vibrational transitions

Figure 8.7 shows the most probable transitions from the $v = 0$ state for the cases where the equilibrium bond length in the upper state is less than, the same as, and greater than that in the lower state. The corresponding spectra are shown underneath. It can be seen that when the two electronic states have very similar r_0 values, the intensity is concentrated mostly in the 0–0 band, but when the r_0 values are quite different, the 0–0 band is quite weak.

If the equilibrium bond length is quite appreciably longer in the upper electronic state than the lower, then the absorption spectrum may show some continuous absorption, as well as the usual line structure. Vertical transitions now meet the curve of the upper state above the dissociation limit; the molecule does not now vibrate, but simply dissociates. The energy of the upper state is no longer quantized, and so absorption of energy can occur over a continuous range of frequencies.

In absorption spectroscopy we may normally ignore transitions which do not originate from the $v = 0$ level, because of the Boltzmann distribution law (see page 95). In emission spectroscopy this is not the case, and then transitions may be observed from many excited vibrational levels, as the molecule is usually not at thermal equilibrium. Nevertheless, the Franck–Condon factors still govern the intrinsic probabilities of the various transitions, and transitions with $\Delta v = 0$ are very strong if the bond lengths in the upper and lower electronic states are the same.

This discussion of intensities may be extended to polyatomic molecules, and vibrational overlap integrals calculated for each normal vibrational mode. However, in practice large changes in bond lengths on excitation are less common, and the $0-0$ transitions often dominate the absorption spectrum.

D. Selection rules

One of the most important aims in calculating intensities by evaluating matrix elements is to predict or explain why many transitions do not appear at all in a spectrum. If it can be shown that R_{ij} for a transition is zero, then that transition will not appear in the spectrum and any generalization about transitions for which R_{ij} is zero will constitute a selection rule. Fortunately it is often much easier to show that a matrix element is zero than to evaluate a non-zero matrix element; by use of symmetry it is not even necessary to know the initial and final wavefunctions exactly.

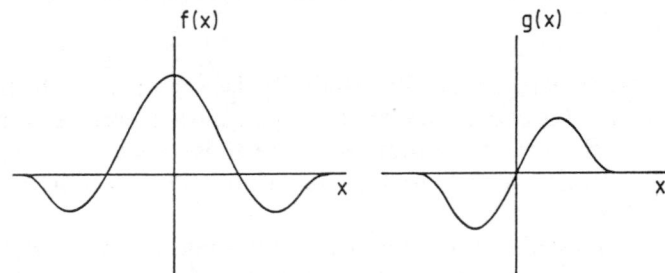

Figure 8.8 Even and odd functions

We can illustrate the principle of this method by considering an example in just one dimension. Figure 8.8 shows the integration of two functions, $f(x)$ and $g(x)$. The function $f(x)$ is said to be an even function, because $f(x) = f(-x)$, whereas $g(x)$ is an odd function, because $g(x) = -g(x)$. Clearly for the even function, $\int_{-\infty}^{\infty} f(x)\mathrm{d}x \neq 0$, whereas for the odd function, $\int_{-\infty}^{+\infty} g(x)\mathrm{d}x = 0$. This may be extended to the calculation of intensities;

the integral $\int_{+\infty}^{+\infty} \Psi_i R \Psi_j d\tau$ will be zero if the function $\Psi_i R \Psi_j$ is odd and non-zero if $\Psi_i R \Psi_j$ is an even function. This in turn will depend on whether each of the functions Ψ_i, R and Ψ_j is odd or even; the multiplication rules are simply even $\times$ even = odd $\times$ odd = even and even $\times$ odd = odd $\times$ even = odd.

As R is a vector, if we change $x \to -x$, $y \to -y$ and $z \to -z$ (considering an example in three dimensions), then $R \to -R$, and so R is odd. Therefore $\Psi_i R \Psi_j$ will be an odd function and the transition i $\to$ j will be forbidden if both Ψ_i and Ψ_j are even functions or if both are odd functions.

This may be applied to the vibrations of an harmonic oscillator, whose wavefunctions were shown in Figure 8.5. The vibrational wavefunctions are even when the quantum number v is even and odd when v is odd. Therefore we may deduce that:

$$\int \Psi_{v=0} R \Psi_{v=1} d\tau \neq 0 \qquad \text{as } \int e.o.o. = \int e \neq 0$$

but

$$\int \Psi_{v=0} R \Psi_{v=2} d\tau = 0 \qquad \text{as } \int e.o.e. = \int o = 0,$$

and so transitions for which $\Delta v = 2, 4, 6$, etc., are forbidden. This conclusion did not require knowledge of the force constant of the bond, nor of the reduced mass of the molecule: the general shape of the wavefunctions was sufficient. It is important to notice here that the converse of our conclusion is not necessarily valid: transitions for which $\Delta v = 1, 3, 5$, etc., do not necessarily have finite intensity. Those integrals may still be zero for some other reason, and in this particular example only transitions for which $\Delta v = 1$ are observed. Nevertheless, this does not contradict our original analysis which indicated that transitions for which Δv was even would not occur in the vibrational spectrum of an harmonic oscillator.

For Raman spectroscopy the integral to be evaluated is now

$$\int \Psi_i \, \alpha \, \Psi_j d\tau,$$

where α is the polarizability. Polarizability is a property which remains unaltered when all the coordinates are reversed, and is therefore even. Thus in Raman spectra the forbidden transitions are those between odd and even vibrational wavefunctions, and we may deduce that the selection rule $\Delta v \neq 1$, 3, 5, etc., applies.

The rotational wavefunctions of a rigid rotator may also be analysed in this way. Figure 8.9 shows the rotational wavefunctions for low values of J and also gives them in analytic form. Just as with the vibrational wavefunctions, we see that rotational wavefunctions are even functions when J is even and odd functions when J is odd. Thus we may deduce in exactly the same way that transitions for which $\Delta J = 2, 4, 6$, etc., are forbidden in absorption spectroscopy, and those for which $\Delta J = 1, 3, 5$, etc., are forbidden in Raman spectroscopy.

The generalization of the use of odd and even functions to examples where the symmetry is more complex lies in the field of group theory, which is beyond

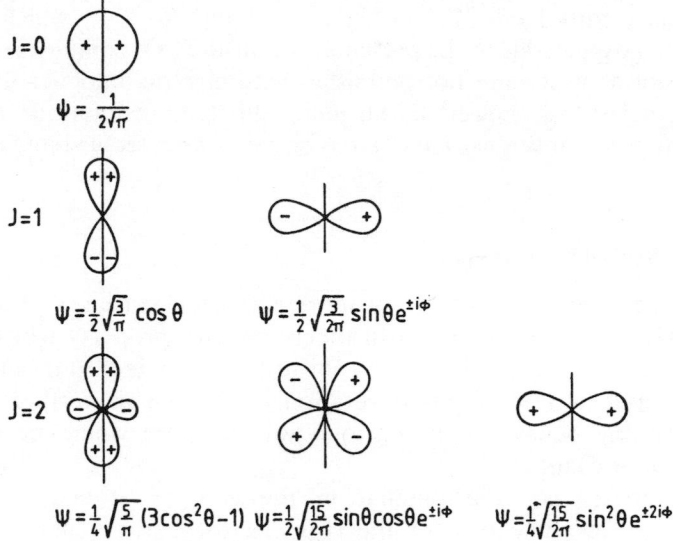

Figure 8.9 Rigid rotor wavefunctions

the range of this book. The principles remain similar, and a transition is forbidden if the direct product of $\Psi_i R \Psi_j$ does not contain the totally symmetric representation.

Breakdown of selection rules

Not infrequently transitions which at first sight ought to be forbidden are actually observed, albeit usually with very feeble intensity. This may be due to a number of causes.

(a) The wavefunctions used in the derivation of the selection rule may have been only approximate. We have already seen that real molecules do not behave exactly as harmonic oscillators, and selection rules based on the harmonic oscillator wavefunctions do not always hold rigidly. Similarly, in electronic wavefunctions it is usual to regard electron spin and orbital angular momentum as entirely independent quantities; in practice spin–orbit coupling, which becomes more important in heavier molecules, prevents either from being precisely defined, and transitions which are formally forbidden may in practice be observed.

(b) The symmetry of the molecule may be altered, permitting a transition which would otherwise be forbidden. One way in which the symmetry of a molecule may be altered is by the application of a strong electric or magnetic field. Alternatively, a new symmetry may be appropriate if molecular collisions are important; in reality we may be observing a system containing a pair of molecules. An example of this is the observation of the formally forbidden $^1\Sigma_g^+ \to {}^3\Sigma_g^-$ transition in O_2, which is seen in liquid oxygen due to the presence of transitory O_4 complexes.

(c) Transitions which are not permitted with electric dipole radiation may nevertheless be allowed as magnetic dipole transitions or as electric quadrupole transitions. These are typically of very feeble intensity.

E.   The effect of nuclear spin

If we consider molecules of two or more equivalent nuclei, such as H_2, N_2, CO_2 or HCCH, then there is a further symmetry property which must be taken into account, and which turns out to have very important implications for the population of different rotational levels. This further symmetry property is the behaviour of the total wavefunction when the equivalent nuclei are interchanged.

This property is more familiar in the case of atomic structure; in constructing wavefunctions for the two electrons in a helium atom, it is important to ensure that the electronic wavefunction is antisymmetric with respect to electron exchange. This prevents the two electrons having the same spin if they occupy the same orbital, which is the generalization known as the Pauli principle.

In the case of molecular structure, the situation is a little more complex. Nuclei with integral spin, such as oxygen and carbon ($I = 0$) or deuterium ($I = 1$), obey Bose–Einstein statistics, according to which the total wavefunction must be symmetric with respect to exchange of nuclei. Nuclei with half-integral spin, such as hydrogen ($I = \frac{1}{2}$) and chlorine ($I = \frac{3}{2}$), obey Fermi–Dirac statistics, according to which the total wavefunction is antisymmetric to exchange of nuclei.

The simplest example of the effect of nuclear statistics on molecular structure is carbon dioxide, where the two oxygen atoms are equivalent. We may consider the effect of nuclear exchange on the electronic, vibrational and rotational wavefunctions separately, and then by multiplication find the effect of nuclear exchange on the total wavefunction. The electronic ground state of CO_2 is $^1\Sigma_g^+$, and this is true for the great majority of stable, centrosymmetric molecules. Exchange of nuclei leaves the electronic wavefunction unchanged in this case. The vibrational wavefunction depends only on the magnitude of the bond length, and this is also unchanged on nuclear exchange. However, the rotational wavefunctions are affected by nuclear exchange: rotational wavefunctions for which J is even are symmetric, but those for which J is odd are antisymmetric with respect to nuclear exchange. For CO_2, in which the oxygen atoms have no nuclear spin, the analysis is now complete, and it can be seen that the total wavefunction must be symmetric to exchange of nuclei, and we are therefore forced to conclude that only rotational states for which J is even can exist.

This remarkable conclusion cannot be tested by looking at the rotational absorption spectrum, as CO_2 has no dipole moment, but both rotational Raman spectra and vibrational spectroscopy confirm that rotational states for which J is odd do not exist. An impressive contrast exists between the rotational Raman spectra of $O^{16}CO^{16}$ and $O^{16}CO^{18}$; in the latter case, the two oxygen nuclei are not equivalent and considerations of Bose–Einstein statistics do not apply. For $O^{16}CO^{18}$ both odd and even values of J are permitted; the spectra are shown in Figure 8.10.

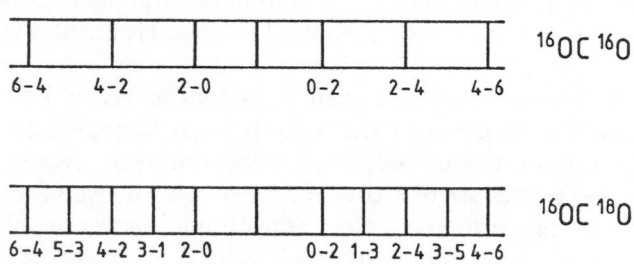

Figure 8.10 Rotational Raman spectrum of CO_2

It might be thought that O_2 would have been a simpler example to choose to illustrate the significance of Bose–Einstein statistics. The argument is very similar in this case, but the electronic wavefunction is now $^3\Sigma_g^-$ rather unusually, and this is antisymmetric with respect to nuclear exchange. Thus for O_2 it is the rotational states with even values of J which do not exist.

Ortho- and Para-hydrogen

To calculate the effect of nuclear statistics on the hydrogen molecule, we must include nuclear spin in the total wavefunction:

$$\Psi = \psi_{el}\,\psi_{vib}\,\psi_{rot}\,\psi_{nuc}$$

As in the case of CO_2, the ground electronic state has symmetry $^1\Sigma_g^+$, and so the electronic and vibrational wavefunctions are left unchanged by exchange of nuclei. Similarly, the rotational wavefunction is symmetric for even values of J and antisymmetric for odd values.

Now a hydrogen nucleus has spin $I = \frac{1}{2}$, and we may write the two possible orientations of this spin as α and β. The possible nuclear spin wavefunctions are then

$\alpha(1)\,\alpha(2)$	$T = 1$	$M_T = 1$
$\beta(1)\,\beta(2)$	$T = 1$	$M_T = -1$
$\alpha(1)\,\beta(2) + \alpha(2)\,\beta(1)$	$T = 1$	$M_T = 0$
$\alpha(1)\,\beta(2) - \alpha(2)\,\beta(1)$	$T = 0$	$M_T = 0$

The first three wavefunctions are those associated with the total resultant nuclear spin $T = 1$ (the degeneracy $2T + 1$ representing the three possible orientations of T) and the last wavefunction is associated with the total nuclear spin $T = 0$. The first three nuclear wavefunctions are clearly symmetrical with respect to exchange of nuclei (1) and (2), whereas the last is antisymmetric.

In order that the total wavefunction shall be antisymmetric to exchange of nuclei, as required by Fermi–Dirac statistics, the rotational levels for which J is even are associated with the total nuclear spin $T = 0$ and those with odd values of J are associated with the total nuclear spin $T = 1$. As the nuclear spin degeneracies are in the ratio of 1:3, it is found that at high temperatures the populations of the even and odd rotational levels in H_2 are also in the ratio of 1:3.

Transitions between odd and even rotational levels of H_2 are strongly forbidden, and if molecules with even values of J are separated from those with odd values, it is found that interconversion does not occur, even over very long periods of time. In the case of hydrogen, the rotational levels are quite widely spaced, and the properties of molecules with odd and even values of J are slightly different. Molecules with odd values of J are referred to as ortho-hydrogen and those with even values of J as para-hydrogen. Other symmetrical molecules exist in ortho and para forms, but because of the closeness of the rotational energy levels, separation is very difficult.

Although interconversion of ortho- and para-hydrogen cannot be achieved by light, or by collisions between hydrogen molecules, interconversion does occur if the atoms can dissociate and then recombine, e.g. on the surface of a catalyst, or if a paramagnetic compound, such as a transition metal ion, produces a strong magnetic field at one nucleus and reorients the nuclear spin. If interconversion occurs at high temperatures, then ortho- and para-hydrogen are produced in the statistical ratio of 3:1, but at very low temperatures, close to absolute zero, all the molecules are formed in the $J = 0$ state and so only para-hydrogen occurs.

In the case of homonuclear diatomic molecules which have nuclei for which $I > \frac{1}{2}$, an analysis similar to that for hydrogen shows that intensity alternation still occurs, with the ratio $(I + 1):I$. This intensity alternation has been used as a method of determining nuclear spin, although it is not very sensitive for large values of I, and it is now more usual to deduce I from the observation of hyperfine structure in atomic spectra or from the fine structure in nuclear magnetic resonance spectra.

Intensity alternation is not only a feature of linear molecules: other symmetrical molecules also exhibit alternation, and the exact nature of the alternation can be used to deduce information about molecular symmetry.

F. Emission of radiation

Not only may molecules absorb light, thereby increasing their energy, but they may also emit light, providing that they are not in their lowest energy state, thereby losing energy. The emission of light may occur in two ways, as shown in Figure 8.11.

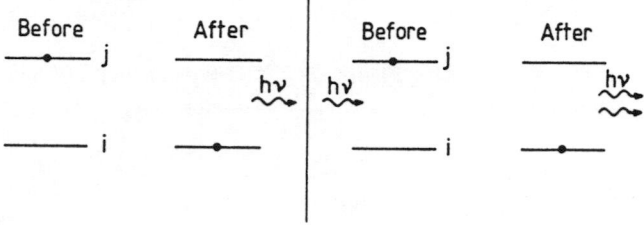

Figure 8.11 Spontaneous and stimulated emission

A molecule in an excited state may spontaneously fall to a lower state, thereby emitting a photon; alternatively, the emission of a photon of frequency v_{ij} may be stimulated by an incoming photon of the same frequency. In this case the emitted photon moves in the same direction as the stimulating photon, and with the same polarization. The rate of spontaneous emission is given by

$$\text{Rate} = A_{ji}\, N_j$$

and that of stimulated emission by

$$\text{Rate} = B_{ji}\, N_j\, \rho(v_{ji}).$$

The proportionality constants are called the Einstein A and B coefficients. Consideration of the frequency distribution in black-body radiation leads to the conclusion that

$$B_{ij} = B_{ji} = \frac{c^3}{8\pi h v^3}\, A_{ji},$$

where B_{ij} is the proportionality constant for the absorption of radiation which was introduced on page 93.

If spontaneous emission is to be an important process, clearly A_{ji} must be large compared with B_{ji}, and this requires v_{ij} to be large. Spontaneous emission is most commonly used in the observation of electronic spectra in the ultraviolet and visible regions and in fluorescence and phosphorescence. Vibrational and rotational spectra, for which v is much smaller, are normally observed in absorption.

G. Linewidths

In most of the spectra that we have encountered so far, individual lines have been drawn as if they were infinitely sharp and occurred at precisely defined wavelengths. In practice this is found not to be the case, and a typical spectral line may appear as in Figure 8.12. In some cases the width of the observed spectral lines may be attributed largely to experimental causes; e.g. the resolving power of a spectrometer may be low. However, even with a perfect experimental set-up, it is still found that spectral lines are not infinitely sharp, and this reflects properties of the energy levels of molecules. There are three main causes of the broadening of spectral lines:

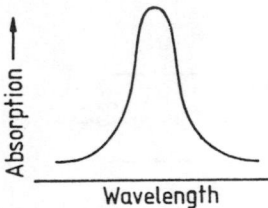

Figure 8.12 The typical shape of a spectral line

(a) Collisions. When a molecule collides with another molecule, its energy levels are modified by the collision and the energies of its transitions are slightly altered. In the gas phase, collisions are relatively infrequent and the broadening is rather small; if wall collisions are unimportant, then the broadening increases with gas pressure. In the liquid phase, collisions are much more important and it is not normally possible to distinguish individual rotational lines.

(b) Doppler broadening. The Doppler effect produces a shift in observed frequency of a wave motion when the source and the observer are in relative motion; it is a familiar effect in sound waves, although it is also used in spectroscopy to estimate the speed of astronomical objects. In a gas there will be a distribution of molecular velocities, with some molecules moving towards an observer and others away. These velocities produce small shifts in the observed frequency of a spectral line, and hence line broadening. In gases the Doppler effect is often greater than the effect of collision broadening, whereas in liquids the reverse is true.

(c) Uncertainty principle. According to Heisenberg's uncertainty principle, the energy of a state is only precisely defined if the lifetime of the state is infinite. If the lifetime of the state is finite, then there will be a corresponding uncertainty in its energy. The lifetime of a ground state may well be essentially infinite, but the lifetime of an excited state will depend on the Einstein A coefficient of spontaneous emission. This limitation, which is not dependent on the precise experimental conditions of temperature, pressure, etc., represents the irreducible minimum linewidth of any spectroscopic transition.

H. Lasers

The spontaneous emission of radiation is a familiar concept; it forms the basis of sodium street lamps and flame tests. In contrast, stimulated emission is much less familiar, and because the Einstein B coefficients are the same for stimulated emission and absorption, absorption is usually the more important process as at thermal equilibrium the lower state will have the greater population. However, if it is possible to arrange for the upper state to have a greater population than the lower state, stimulated emission predominates.

Stimulated emission is the essential prerequisite for laser action. We need a greater population in the upper level than in the lower and this is achieved by 'pumping'. Two alternative approaches are employed. In pulsed lasers the species whose energy levels are involved in laser action is an ion of perhaps chromium or europium. These sit as impurities in a lattice of ruby. The atoms and the crystal are excited to higher electronic levels by a short pulse of light. The excited crystal energy is then used in promoting the impurity ion into an excited state so that momentarily there is a population inversion. Alternatively, in continuous working lasers, one type of atom, perhaps a helium atom, is excited to an upper state by an electric discharge. These excited helium atoms then bump into and transfer their energy to neon atoms, giving excited states. Whilst the latter may be few in number by comparison with the ground state they are far more numerous than the populations of some others excited but lower lying levels (as in Figure 8.13). We then again have a population inversion, ripe for the stimulated emission of radiation to predominate.

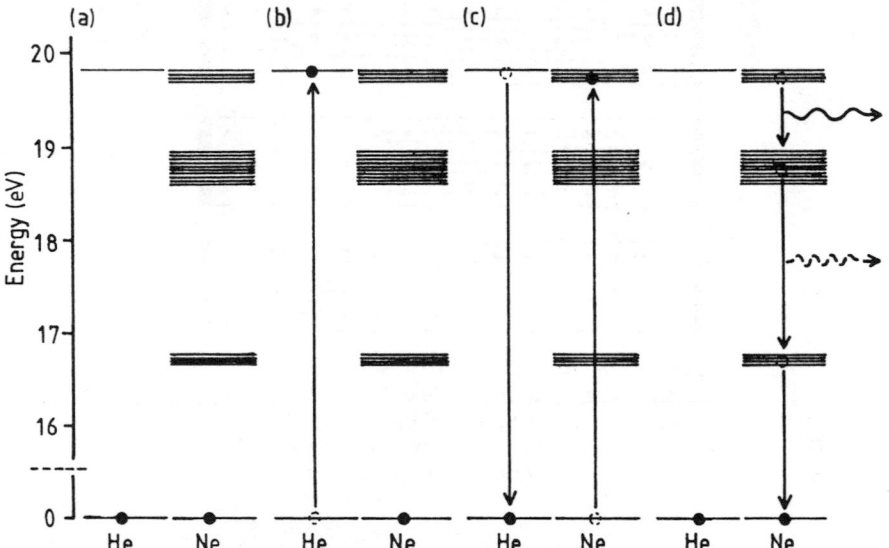

Figure 8.13 The excitation of stimulated emission in the helium–neon laser: (a) before excitation, (b) He excited, (c) neon excited, (d) laser action

This stimulated radiation may be amplified by putting the medium between parallel reflectors. Light will build up between the mirrors as shown in Figure 8.14, giving light perpendicular to the mirrors, as well as being in phase and monochromatic since only exact numbers of wavelengths of the light will fit into the box and rebound without destructive interference.

This monochromatic coherent light once amplified is called LASER light (Light Amplification by Stimulated Emission of Radiation). A beam may be extracted by having one of the reflecting mirrors partially transparent.

The uses of this light are now very numerous, but not least to spectroscopists the laser has become a wonderful tool for beautiful, exact and detailed experiments.

110

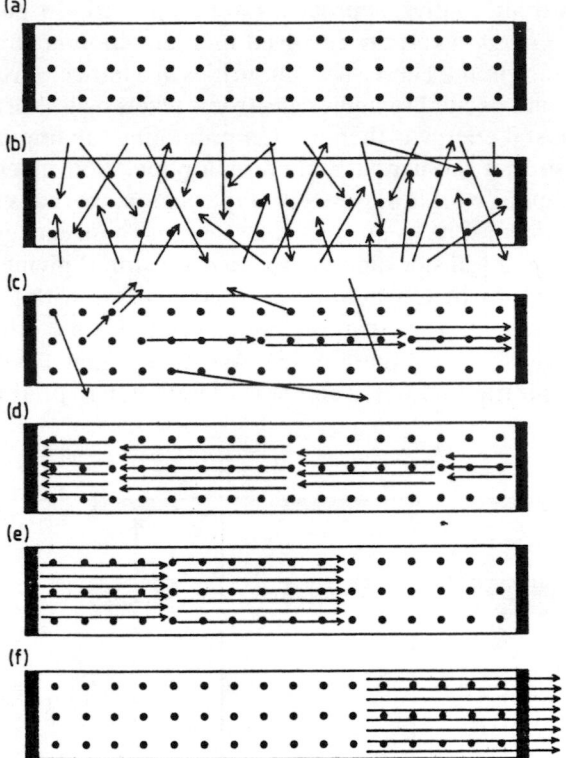

Figure 8.14 Amplification of stimulated emission: (a) ground state atoms, (b) excitation, (c) mixture of spontaneous and stimulated emission, (d), (e) build up of stimulated emission, (f) transmission of stimulated emission

I.   Summary

The intensities of spectral lines depend on the intrinsic probability associated with a transition and on the population of the appropriate energy level.

Intrinsic probabilities in turn depend on the wavefunctions of the two levels concerned in the transition; in some cases the symmetry properties of the wavefunctions can be used to show that the intrinsic intensity is zero, and selection rules can be obtained.

Populations of energy levels at equilibrium are given by the Boltzmann distribution law. Many rotational states may be occupied, but only the lowest vibrational and electronic states have significant populations under most conditions. Non-equilibrium conditions may permit stimulated emission to dominate over the more normal spontaneous emission, and create the possibility of laser action.

J. Problems

1. Find the molar absorption coefficient given the following data for the absorption of light by a sample of bromine in a cell of path length 0.2 cm.

$[Br_2]/mol\ dm^{-3}$	0.001	0.005	0.010	0.050
Transmission per cent	81.4	35.6	12.7	3×10^{-3}

2. What are the proportion of halogen molecules in the first excited electronic state at 500 K (ω_e for F_2, Cl_2, Br_2 and I_2 are 892, 565, 323 and 215 cm^{-1} respectively).
3. The molecule BF has a rotational constant 1.52 cm^{-1}. Which rotational states will be most populated at 500 K?
4. What differences would there be between the pure rotational Raman spectra of H_2 and D_2 (For H, $I = \frac{1}{2}$ and $m = 1$; for D, $I = 1$ and $m = 2$)?
5. The hydrated Mn^{2+} ion has a sextuplet ground state, with five unpaired electrons. Why is the ion almost colourless?

Chapter 9

Photoelectron spectroscopy and the structure of ions

A. Introduction

We have seen that photons in the microwave and infrared regions of the electromagnetic spectrum are sufficiently energetic to alter the rotational and vibrational energies of a molecule, and that visible and ultraviolet photons may excite an electron from one orbital to another. If we supply even larger amounts of energy to a molecule, an electron may be removed from it altogether, leaving behind a positive ion. This chapter is concerned with forms of spectroscopy in which ionization occurs; the use of photons as an energy source forms the basis of photoelectron spectroscopy, and electron bombardment the basis of mass spectrometry.

These experiments are concerned with the measurement of the ionization energies of the electrons in a molecule; no selection rules are involved and so it is possible to obtain a measure of the binding energy of every electron. Not only does this provide a unique tool for investigating the electronic structure of a molecule, but it also has very important applications in analytical chemistry.

There is one important difference between these experiments and the familiar absorption experiments that we have studied in earlier chapters. When ionization occurs, the translational energy of the ejected electron is not quantized and therefore there is no single photon or electron energy which is required to produce ionization: any energy in excess of the ionization will suffice. It is not possible therefore simply to irradiate the sample with photons or electrons of varying energies, and instead a monochromatic beam is used; the energies of the ejected electrons are then measured experimentally. The ionization energies of the molecule may then be obtained from the law of conservation of energy:

$$h\nu = \text{ionization energy} + \text{electron kinetic energy}.$$

The recoil of energy of the ion may be neglected as its mass is much greater than that of the photoelectron.

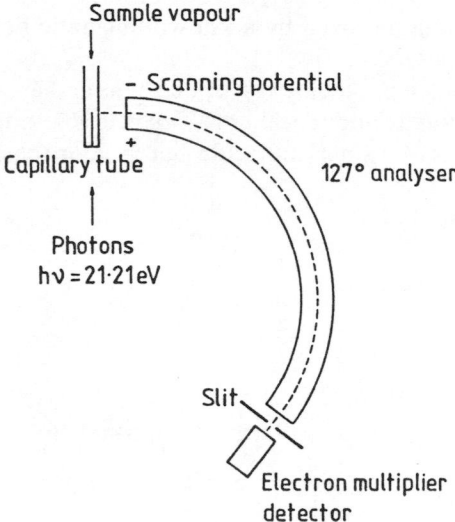

Figure 9.1 Schematic view of an ultraviolet
photoelectron spectrometer

B. Ultraviolet photoelectron spectroscopy

The ionization energies of the valence electrons of molecules fall in the range 5 to 30 eV. If these are to be studied by photoelectron spectroscopy, the exciting photon must fall in the vacuum ultraviolet part of the spectrum. Many discharge lamps produce photons in this region, but the problem is to find a source which is inherently monochromatic, as the use of diffraction gratings as wavelength selectors produces a large loss in intensity. Most of the studies which have been performed to date have used a helium discharge lamp, which emits a single line at 58.4 nm (21.2 eV), although a number of other sources have been used, particularly another helium line at 40.8 eV.

Although solids and liquids may be studied in ultraviolet photoelectron spectroscopy, gas phase samples have been most widely used. This produces two problems: there is no suitable material which can form a window for vacuum ultraviolet photons, separating the lamp from the sample, and the sample pressure which is needed to produce a detectable number of photoelectrons is greater than the pressure which can be allowed in the electron energy analyser, where electron–molecule collisions must be avoided. These two problems are overcome by differential pumping, with the lamp and the electron energy analyser being connected to the sample chamber by capillary tubes. The sample is slowly bled into the sample chamber, and the lamp and the analyser are separately pumped to a much lower pressure. A schematic diagram showing an ultraviolet photoelectron spectrometer is given in Figure 9.1. The kinetic energy of the photoelectrons may be found either by

114

measuring the deflection caused by a known magnetic field or by the use of electrostatic potentials.

A typical ultraviolet photoelectron spectrometer has a resolution of the order of 0.01 eV; this is about 100 cm^{-1}, which is less than the spacing of the vibrational levels of a small molecule but greater than the spacing of the rotational levels.

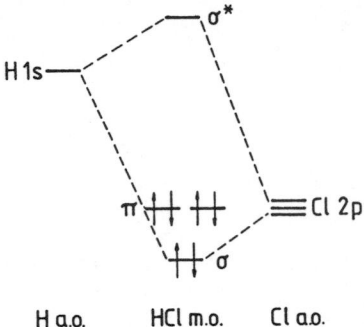

Figure 9.2 Molecular orbital diagram for HCl

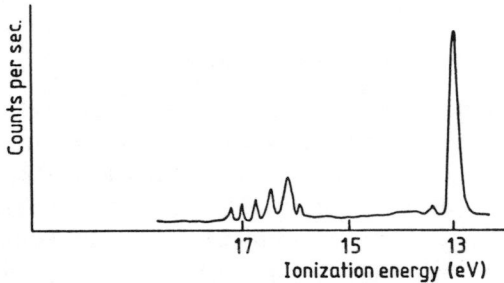

Figure 9.3 The ultraviolet photoelectron spectrum of HCl

Figure 9.2 shows the molecular orbital diagram for the valence orbitals of hydrogen chloride, illustrating the formation of a single bond between the 1s orbital of the hydrogen atom and the 2p electron of the chlorine atom. Figure 9.3 shows the experimental photoelectron spectrum of hydrogen chloride; it can be seen that there are two distinct bands, one at about 13 eV and the other at about 16 eV. The first band corresponds to the removal of a non-bonding π electron and the second band to the removal of a bonding σ electron. Remember that in a photoelectron spectrum we are always observing processes of the type $S \rightarrow S^+$; S^{2+} ions are not formed in this experiment and so second ionization energies, corresponding to the process $S^+ \rightarrow S^{2+}$, are not determined.

Clearly there is a close relationship between the appearance of the molecular orbital diagram in Figure 9.2 and the photoelectron spectrum in Figure 9.3; it has been said that photoelectron spectroscopy provides the most direct evidence in favour of molecular orbital theory. In this example the analysis of the spectrum is straightforward; the experiment merely confirms elementary ideas about the way in which covalent bonds are formed. However, in more complex examples it may be much more difficult to assign the bands unambiguously, and molecular orbital calculations have been used widely to assist in interpretation. Comparison with chemically similar molecules can also play an important role in assignment, and information may also be available from the appearance of fine structure in a band, as we shall see below.

C. Vibrational fine structure

It is quite common to find that a band in a photoelectron spectrum corresponding to the removal of a particular electron will show fine structure when examined at high resolution. If we are to understand how this fine structure arises, we shall need to keep in mind the idea that a photoelectron spectrum does not simply measure the energy of an electron in a neutral molecule, but the difference in energy between the neutral molecule and its positive ion. Just as in infrared spectroscopy, we may normally assume that the molecule starts in the lowest vibrational state of the lowest electronic state, but if the ion has several possible states to which transitions may occur, then fine structure will appear in the photoelectron spectrum.

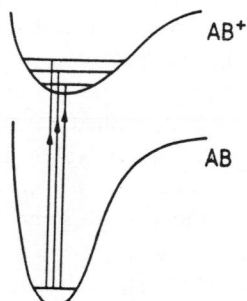

Figure 9.4 Ionization to excited vibrational levels of an ion

The most important cause of fine structure in photoelectron spectroscopy is the vibrational motion of the ion. Figure 9.4 shows the potential energy curves for a typical diatomic molecule and its ion. Transitions originate from the $v = 0$ level of the molecule, but the ion may be formed in a number of different vibrational states, and so a series of lines, roughly equally spaced, appears in the photoelectron spectrum. The separation of these lines depends on the vibrational energy spacing of the ion, and not of the neutral molecule. The energy difference between the $v = 0$ levels of the molecule and ion is known as the adiabatic ionization energy, whereas the energy of the most probable transition is known as the vertical ionization energy.

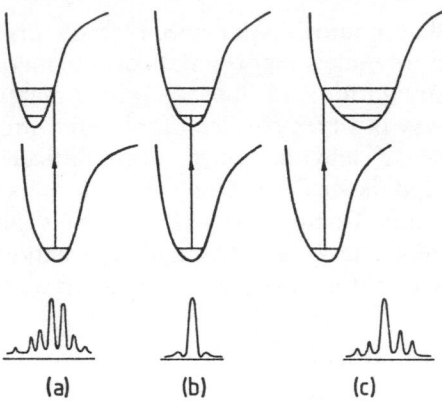

Figure 9.5 Ionization of (a) an antibonding electron, (b) a non-bonding electron and (c) a bonding electron

The intensities of vibrational lines in a photoelectron spectrum are governed by the Franck-Condon principle, just as in electronic absorption spectra. The removal of a bonding electron from a molecule produces an increase in bond length, the removal of an antibonding electron produces a decrease in bond length and the removal of a non-bonding electron has little effect. These three cases are shown in Figure 9.5. When a non-bonding electron is removed from a molecule, the intensity is therefore concentrated in the $0-0$ line, and the vibrational fine structure is not very pronounced. However, when a bonding or an antibonding electron is removed, the vibrational structure is much more pronounced and a number of strong lines are observed. The vibrational fine structure may therefore be of great value in assigning the bands in a photoelectron spectrum. Further assistance may come from measurement of the vibrational frequency of the ion; in general the removal of a bonding electron produces a broader, shallower curve, and so the vibrational frequency of the ion is less than that of the corresponding molecule. When an antibonding electron is removed, the reverse is true.

These principles are well illustrated by the photoelectron spectrum of HBr, which is shown in Figure 9.6 along with the corresponding molecular orbital diagram. The band at 12 eV represents the removal of a π electron, which can be seen from the molecular orbital diagram to be non-bonding. This band shows very little vibrational structure (the two distant peaks do not represent the vibrational structure, but are caused by spin–orbit coupling, as we shall see below). The band at 15 eV represents the removal of the bonding σ electron, which gives rise to appreciable vibrational structure. Table 9.1 shows the vibrational frequencies of HBr and HBr$^+$, which further confirm the bonding and non-bonding nature of the electrons.

The resolution of ultraviolet photoelectron spectroscopy is such that individual rotational lines are not observed. However, in some cases it is

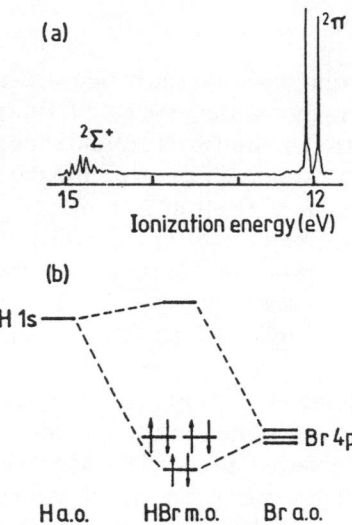

Figure 9.6 (a) Schematic photo-electron spectrum of HBr. (b) The corresponding molecular orbital diagram

Table 9.1   Vibrational frequencies of HBr and HBr$^+$

Species	Frequency (cm^{-1})	Electron removed
HBr	2560	
HBr$^+$, $^2\Pi$	2420	Non-bonding
HBr$^+$, $^2\Sigma^+$	1290	Bonding

possible to observe an asymmetry in the spectrum of a vibrational transition due to rotational structure, even though no individual lines may be observed. Just as in electronic spectroscopy, the rotational lines will form a band-head if the rotational constants, and hence the bond lengths, of the molecule and the ion are appreciably different (compare page 81). If the molecule has a greater bond length than the ion, i.e. if an antibonding electron is being removed, then the R branch shows a band-head and the intensity slowly tails off to lower energies. If the ion has a greater bond length than the molecule, then the P branch shows a band-head and the intensity slow tails off to higher energies. In high resolution photoelectron spectra these two cases may sometimes be distinguished, even though only the rotational envelope may be observed; this may provide further help in assigning photoelectron bands.

D. Spin–orbit coupling

A photoelectron spectrum may also show fine structure caused by spin–orbit coupling. Again we may choose the molecule HBr as our example. Hydrogen bromide has all its electrons paired in its ground state, so it has no resultant spin or orbital angular momentum. This is true for the great majority of simple molecules. However, if a π electron is removed, forming HBr^+, the ion now has spin angular momentum of $\frac{1}{2}$ due to its one unpaired electron and orbital angular momentum of 1 due to its singly occupied π orbital. (Actually the angular momementa are $\sqrt{\frac{1}{2}(\frac{1}{2} + 1)}$ and $\sqrt{\frac{3}{2}(\frac{3}{2} + 1)}$ respectively, and are measured here in units of $h/2\pi$. This does not affect the argument in any way.)

The total angular momentum must be $(1 + \frac{1}{2}) = \frac{3}{2}$ or $(1 - \frac{1}{2}) = \frac{1}{2}$, depending on the relative orientations of spin and orbital angular momenta. These two states are referred to as $^2\Pi_{3/2}$ and $^2\Pi_{1/2}$ respectively, and their energies are not exactly identical. The result is that two peaks occur in the photoelectron spectrum of HBr, which was shown in Figure 9.6. The magnitude of this spin–orbit splitting varies considerably from element to element and increases with atomic number. For the $^2\Pi$ states of the hydrogen halides the values are: HCl^+ 0.08, HBr^+ 0.33 and HI^+ 0.66 eV. When an electron is removed from an orbital with no angular momentum, such as the orbital in HBr, the resulting ion has no orbital angular momentum and so there is no spin–orbit splitting. This effect may therefore be useful in the assignment of bands in a spectrum; in larger molecules, it may help in the identification of bands caused by the removal of electrons which are localized on heavy atoms.

E. Koopmans' theorem

We have already referred to the close connection between molecular and orbital diagrams and the appearance of the photoelectron spectra. Theoretical molecular orbital calculations have now become routine for small and medium-sized molecules, and these give the electronic wavefunction for each electron, ϕ_i, and the one-electron energy for each molecular orbital, ε_i. (There are some approximations involved for all other than the smallest molecules, but these need not concern us here.) It has therefore become commonplace for photoelectron spectra and molecular orbital calculations to be considered together, with each being interpreted in the light of the other.

This procedure is not, however, strictly rigorous, and it should be treated with caution. Its theoretical basis lies in the theorem due to Koopmans, who found that if all the orbitals in a molecule remain exactly the same when ionization occurs, then

Ionization energy $= \varepsilon_i$.

This is not a very realistic model, as the shape and size of a given molecular orbital depend on the balance of two factors, the electrostatic attraction of the nuclei and the repulsion of the other electrons. When one of the electrons is removed the electron repulsion term changes, and so all the molecular orbitals will alter somewhat. The orbitals are said to relax, the energy involved being called relaxation energy. As a result Koopmans' theorem overestimates ionization energies; Table 9.2 compares the ionization energies and orbital energies of carbon monoxide.

Table 9.2   The ionization and orbital energies of carbon monoxide

Orbital	$\sigma(2p)$	$\pi(2p)$	$\sigma(2s)$
Orbital energy (eV)	15.09	17.40	21.87
Ionization energy (eV)	14.01	16.91	19.72

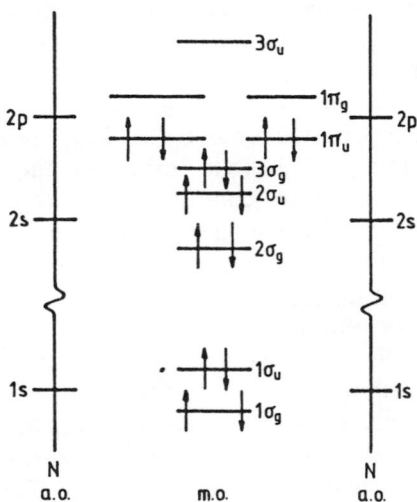

Figure 9.7 The occupied molecular orbitals in the ground electronic state of N_2

In many cases the relaxation energy is not very important, and orbital energies can be of great value in assigning spectra. However, there are dangers inherent in the process as the relaxation energies may differ from one orbital to another. This can be illustrated by the case of N_2. Many molecular orbital calculations have shown that the ordering of orbitals is as shown in Figure 9.7. On the basis of Koopmans' theorem we should expect a π_u electron to be most easily removed, and therefore the ground state of N_2^+ should be $^2\Pi_u$. But the electronic spectrum of N_2^+, which is well known to astrophysicists, and the

120

photoelectron spectrum of N_2 both confirm that the ground state of N_2^+ is $^2\Sigma_g^+$. This does not involve us in any dilemmas; we may write the configurations of N_2 and N_2^+ as:

$$N_2: \ldots\ldots \sigma_g^2 \, \pi_u^4$$

$$N_2^+: \ldots\ldots \pi_u^4 \, \sigma_g^1$$

where the orbitals appear in order of energy. Although the π_u orbitals are similar in N_2 and N_2^+, they are not exactly the same functions, and the ordering of the π_u and σ_g orbitals alters when an electron is removed. More technically, there are other small approximations in Koopmans' theorem due to the neglect of correlation energy with which we need not be concerned here.

F. Applications of ultraviolet photoelectron spectroscopy

Photoelectron spectroscopy is still a relatively new experimental technique, but its rapid progress over the last fifteen years suggests that it may become established as one of the major techniques for the investigation of molecular structure. Inevitably early studies concentrated on molecules whose electronic structures were already well established, but more recently experimenters have used photoelectron spectroscopy as a method of solving genuine chemical problems.

Photoelectron studies on small molecules have provided a wealth of information on the properties of positive ions, both in their ground and excited states. Here photoelectron spectroscopy has often been used in conjunction with other techniques, such as vacuum ultraviolet absorption spectroscopy and mass spectroscopy; the fragmentation processes of molecular ions have been clarified by photoelectron spectra.

For larger molecules studies have concentrated on the binding energies of individual molecular orbitals. Variations in these binding energies have been used to provide information on inductive effects, delocalization of electrons, through space interactions between formally non-bonding orbitals, aromaticity, $p \rightarrow d\pi$ bonding, the bonding in low valence transition metal complexes and the geometries of molecules and ions.

Ultraviolet photoelectron spectroscopy has also proved of some value as an analytical tool; transient free radicals may be detected and simple features of the photoelectron spectra of complex molecules may allow identification of functional groups. Photoelectron spectroscopy can also be a useful tool for distinguishing between closely related isomers. Although most studies have been conducted on gases, there have been some solid state studies. Ultraviolet photons are unable to penetrate solids very deeply, but they have been used to give information on the valence bands of solids and on the chemical nature of gases absorbed on their surfaces.

G. X-ray photoelectron spectroscopy (ESCA)

Core electrons have binding energies in the region 100 to 10 000 eV, and these electrons may not be removed by photons in the vacuum ultraviolet. However, they are removed by X-rays, and X-ray photoelectron spectroscopy has proved to be a powerful technique in chemical analysis. It is often referred to as Electron Spectroscopy for Chemical Analysis (ESCA). It might be imagined that X-rays could also be used for the investigation of the ionization energies of valence electrons, but as the resolution in ESCA is generally only about 1 eV, the fine structure which is an important feature of ultraviolet photoelectron spectra would be lost; ESCA also has the disadvantage of being a more expensive technique.

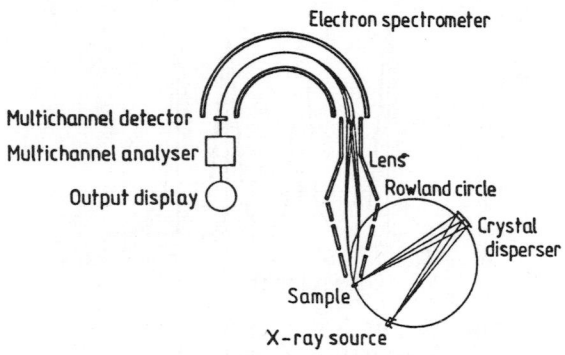

Figure 9.8 Schematic X-ray photoelectron spectrometer using monochromatized X-rays as the source

Figure 9.8 shows a typical arrangement for the observation of X-ray photoelectron spectra of a solid sample. Gases have been used, and give rather sharper spectra, but they tend to give rather low photoelectron intensities. X-rays are generated by directing a beam of high energy electrons at a metallic target; these eject core electrons and a transition then occurs in which another electron falls to replace the core electron, emitting an X-ray. The most intense line comes from a $2p$ electron falling to the $1s$ shell, and this gives rise to the potassium line. Magnesium and aluminium are commonly used as target materials, but heavy atoms can be used to generate higher energy X-rays. The X-ray source is separated from the sample by a thin metallic window, so differential pumping is not necessary.

The energies of the valence electrons of the atom undergo appreciable alteration when a molecule is formed, but the core electrons remain largely unchanged and their energies are shifted little from the free atom values. Table 9.3 gives the ionization energies of the $1s$ electrons of the elements of the first and second short period; they increase steadily with atomic number, roughly following Moseley's relationship, $E \propto Z^2$. These values are sufficiently separated from each other for the appearance of the ESCA spectra to reveal the elements present in a given sample, making ESCA a useful technique in chemical analysis.

122

Table 9.3   Ionization energies.

Atomic number	3	4	5	6	7	8	9
Ionization energy (eV)	55	111	188	285	399	532	686

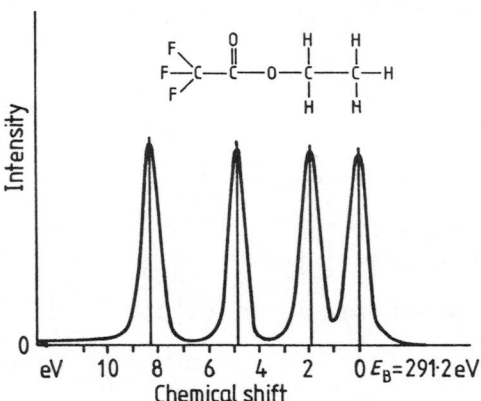

Figure 9.9 The X-ray photoelectron spectrum of $CF_3COOCH_2CH_3$

However, the real power of ESCA lies in the fact that chemical shifts may also be observed. Thus although carbon atoms give rise to ESCA peaks close to 285 eV, the precise energy of a peak depends on the chemical environment of the corresponding atom. Figure 9.9 shows the ESCA spectrum of ethyl trifluoroethanoate, $CF_3COOCH_2CH_3$. The interpretation of this spectrum is straightforward, and may be confirmed by comparison with the spectra of related molecules such as $CH_3COOCH_2CH_3$ and CF_3COOCH_3. The peak at a shift of 8 eV corresponds to the carbon atom in the CF_3 group, that at 5 eV to the carbon in the $C=O$ group and the doublet near zero to the carbon atoms in the CH_2 and CH_3 groups. This pattern may be rationalized simply: the fluorine atoms are strongly electron-withdrawing, and therefore the electron density round the CF_3 carbon is lowest and its $1s$ electron is the most difficult to remove. The oxygen atoms are a little less electron-withdrawing, so the electron density round the $C=O$ carbon is a little higher and the $1s$ electron is a little easier to remove. In this particular example, it is even possible to distinguish between the electron densities on the CH_2 and CH_3 groups.

The relationships between ESCA shifts and the charges on individual atoms in molecules have been investigated in many cases, and in general good correlations have been found, particularly if the effect of neighbouring atoms is taken into account. Charges may be estimated from electronegativity values or taken from molecular orbital calculations.

The observation of chemical shifts has made the ESCA a very powerful technique for the investigation of molecular structure. Thus the structure of cysteine disulphur dioxide was investigated by Siegbahn; two possible structures had been suggested, R–S–S–R and R–S–S–R. The ESCA spectrum showed that there were two different peaks due to sulphur, which ruled out the symmetrical structure and established R–S–S–R as the correct structure. In other examples information may be obtained from the intensities of ESCA lines; for a given element, the intensity is very closely proportional to the number of atoms, and so an ESCA spectrum reveals not only the number of different chemical environments of each element but also how many atoms are found in each environment.

The attraction of a technique which can probe the local environments of the nuclei of all elements in the Periodic Table regardless of nuclear spin prompted the belief that ESCA would prove as versatile and important as nuclear magnetic resonance. This hope has largely remained unfulfilled. The major defect of ESCA is its rather low resolution, which prevents the observation of either fine structure or small chemical shifts. In principle this problem may be overcome by monochromatizing the exciting radiation; if aluminium is used as an X-ray source the linewidth is about 1 eV, but for heavier atoms 10 eV is not uncommon. The use of a monochromatizer helps to sharpen individual ESCA lines, but also produces a marked drop in intensity, which in many cases means that there is little net advantage.

Perhaps the best hope for an intense source of monochromatic X-rays is synchrotron or storage ring radiation. Particle physicists explore the nature of subatomic particles by collision experiments which require high energy particles. Electrons are accelerated round a large diameter circle, and while travelling in this way emit light whose frequency depends on their velocity and radius. The type of source favoured by particle physicists happily emits X-rays which could be of great value to structural chemists. Whether these sources will become readily available to chemists is, of course, another question.

H. The Auger effect

We have considered so far only the primary step which occurs when a high energy photon collides with a molecule, namely photoelectron emission. However, secondary steps may follow, and these can also result in further electron emission, which produces spurious peaks in the photoelectron spectrum. Figure 9.10 shows the possible fates of excited ion S^{*+}. The ion which is produced in the initial step has one of its core orbitals incompletely

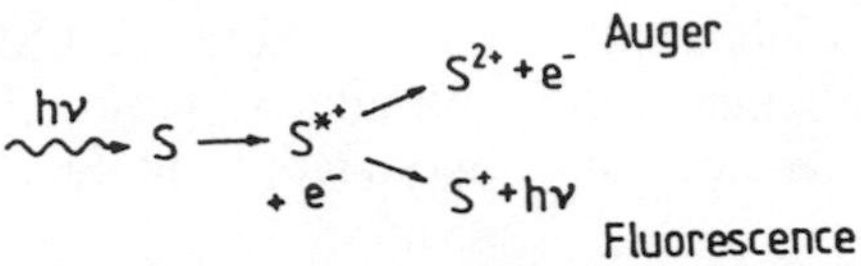

124

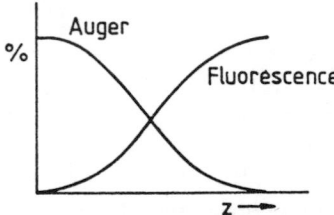

Figure 9.10 The Auger effect

filled, and another electron may fall into this vacancy. The energy released in this step may appear as light, in which case the process is called fluorescence, or alternatively it may be used to eject another electron, which is the Auger effect.

The Auger effect can cause problems in the interpretation of X-ray photoelectron spectra, as the Auger electrons are analysed along with the ordinary photoelectrons and thus give rise to extra peaks. This may be overcome by altering the wavelength of the exciting X-rays which produces a change in the photoelectron energies, as electron kinetic energy $= h\nu -$ ionization energy. However, the energies of the Auger electrons remain unchanged, as their energy is a property only of the levels of the S^+ ion from which they are emitted.

Figure 9.11 Relative amounts of Auger and fluorescent emission as a function of atomic number

Study of the Auger effect is an important form of spectroscopy in its own right; it differs from photoelectron spectroscopy in that it does not require a monochromatic light source. It is most widely used for studying atoms of low mass; as Figure 9.11 shows, fluorescence is a much more likely process for heavy atoms. Although Auger spectra can be obtained by X-rays, the use of electron beams is more usual.

The Auger effect may be used like ESCA to identify the atoms present in a sample. Chemical shifts may also be observed, but the situation is more complicated because there are two energy levels of the S^+ ion involved in electron emission. If the upper level is a valence orbital there is no simple correlation between atomic charges and chemical shifts. Auger studies have been conducted largely on solid samples; the technique is restricted largely to the observation of surfaces. Nevertheless, Auger spectroscopy has proved an extremely sensitive and very important method of studying surface contaminants.

I. Summary

Photoelectron spectroscopy leads to information about the difference in energy between ground state molecules and their ions in a variety of electronic states. The technique measures ionization energies, and thus provides a direct picture of electronic energy levels. The vibrational structure gives information of vibrations in the ion and on changes of geometry on ionization; it may be used to distinguish non-bonding from bonding and antibonding electrons.

When the exciting radiation is in the form of monochromatic X-rays, the ionization energies of core electrons can be measured. Shifts are discernable between the ionization potentials of $1s$ electrons of a given nucleus, depending on its chemical environment.

J. Problems

1. In a PES experiment light of wavelength 58.4 nm is used, and it is found that CO emits electrons with kinetic energy 7.2 eV. Using the conversion 1 eV $= 8065$ cm^{-1}, calculate the ionization energy of the CO molecule.

2. The m.o. diagram of O_2 was given in Chapter 3. The ground state of O_2 is $^3\Sigma_g^-$; what is the symbol for the state of the O_2^+ ion formed by removal of the electron with the lowest ionization energy? Would you expect the band in the p.e.s. due to this ionization to show extensive vibrational structure?

3. In the photoelectron spectrum of Ar two peaks are observed very close in energy. What is the likely origin of this?

4. The photoelectron spectrum of N_2 is given below: how can it be interpreted?

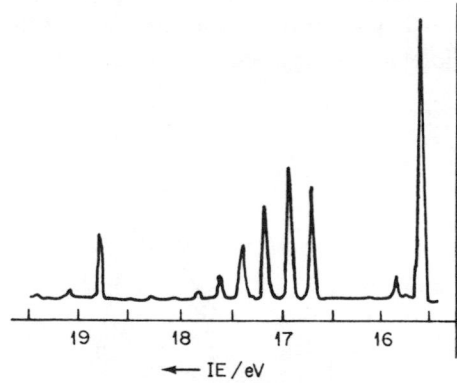

5. Predict the X-ray photoelectron spectrum of the azide ion N_3^- in the region of ionization of nitrogen $1s$ electrons.

Chapter 10

Electronic spectra of condensed phases

A. Introduction

In Chapter 7 we saw that most molecules absorb light in the visible and ultraviolet regions of the spectrum, and that this causes electronic transitions in the molecule. Changes in rotational and vibrational quantum numbers are also possible, and the spectra of gas phase samples therefore consist of vibrational bands, which at high resolution are seen to consist of closely packed lines.

Molecules in condensed phases are constantly bombarded by other molecules, causing changes in their rotational and vibrational energies. Their spectra do not show lines due to individual rotational transitions, as the lines are broadened so much that they overlap with neighbouring lines, and so absorption bands with no rotational structure are observed. In some cases it is possible to observe vibrational structure, but in others, particularly in polar solvents, even this is lost, and a single broad peak is obtained. Figure 10.1 shows the ultraviolet absorption spectrum of ethanal, dissolved in heptane and in ethanol.

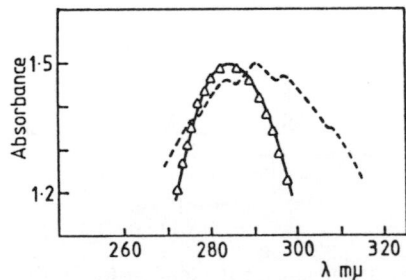

Figure 10.1 Spectrum of ethanal in heptane and ethanol (ethanol with no vibrational structure)

126

It might be thought that electronic spectroscopy in condensed phases is therefore of rather limited value; in fact it has been used very widely and gives valuable information to analytical chemists and those interested in molecular structure alike. This is because most electronic transitions are localized in a particular bond in a molecule, and so many functional groups give rise to absorption bands at characteristic frequencies. However, the precise frequencies do depend to some extent on the structure of the rest of the molecule, and so electronic spectra can give information on the various groups in a molecule and on their relative positions. As spectra can be obtained rapidly using only a very small sample, their use has become routine, particularly in organic chemistry.

Many molecules have now been studied in the visible and ultraviolet regions, and tables which predict absorption frequencies constructed. Table 10.1 gives some typical examples.

Table 10.1

Characteristic group bands	
C=C	180 nm
C=O	280 nm
C=S	400 nm
C−Cl	170 nm
C−I	260 nm
Effect of substituents on C=C absorption	
−OR	+30 nm
−Cl	+5 nm
−CH$_3$	+5 nm

Light with wavelengths from 800 to 400 nm lies in the visible region of the spectrum; shorter wavelength light lies in the ultraviolet. Routine spectral measurements go down to about 200 nm, after which absorption by O_2 molecules interferes; this range can be extended to about 180 nm by flushing the spectrometer with N_2 gas. Measurements at wavelengths shorter than 180 nm require the system to be evacuated (the 'vacuum ultraviolet') and are generally of less value.

B. Spectra of organic molecules

Alkanes

The alkanes absorb only at very short wavelengths in the vacuum ultraviolet part of the spectrum. Transitions are observed at about 130 nm; these are usually rather featureless and have not been widely used. The alkanes contain only single bonds, i.e. σ electrons, and the transitions are from a bonding σ

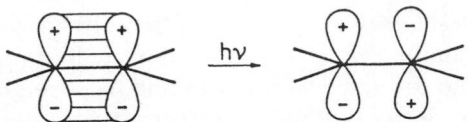

Figure 10.2 The (π^*, π) transition in ethene

orbital to an antibonding σ^* orbital; these are described as (σ^*, σ) transitions. Cyclopropanes are an exception to this rule and absorb at longer wavelengths (about 190 nm). This unusual behaviour may be attributed to ring strain, which destabilizes the bonding orbitals.

Alkenes

Alkenes give rise to an absorption band at about 190 nm; this absorption is due to the promotion of a bonding π electron in the double bond to an antibonding π^* orbital. This is a (π^*, π) transition, and is shown in Figure 10.2. The variation of this absorption on substitution has been extensively studied, and measurement of this band can give precise structural information. In the excited state the π–bond has been broken and the barrier to rotation about the double bond is removed. The absorption of radiation can be followed by rotation, and so in substituted alkenes a different isomer can be produced if the double bond is later reformed. Note that the excitation of σ electrons requires considerably more energy and is not usually observed.

Conjugated dienes

If a molecule contains two double bonds which are well separated from each other, each gives rise to an absorption at about 190 nm independently of the other. However, if the two double bonds are adjacent to each other, as in $CH_2=CH-CH=CH_2$, their π orbitals overlap and the system is said to be conjugated. The (π^*, π) excitation energy is now lower and the band moves to longer wavelengths. In the LCAO approximation the four π orbitals overlap, giving molecular orbitals as shown in Figure 10.3. The four electrons occupy the bonding orbitals, ϕ_1 and ϕ_2, and the observed transition is the excitation of an electron from ϕ_2 to ϕ_3.

The interaction of the two double bonds can be represented in a molecular orbital diagram as in Figure 10.4, which shows how the excitation energy decreases on conjugation.

If the conjugation is lengthened to include more double bonds then the (π^*, π) transition moves to still lower energies and eventually falls into the visible part of the spectrum. Many natural products have long conjugated structures, and this (π^*, π) transition is the reason why carotenes, and hence carrots, are coloured.

The effect of conjugation on electronic spectra has been used to obtain conformational information in the biphenyls. The spectra of $C_6H_5-C_6H_5$ and of $CH_3C_6H_4-C_6H_4CH_3$ resemble each other, provided that the $-CH_3$ groups occupy the m or p-positions in the rings. However, if the two $-CH_3$ groups are both in the o-positions, then the spectrum changes markedly. This is because

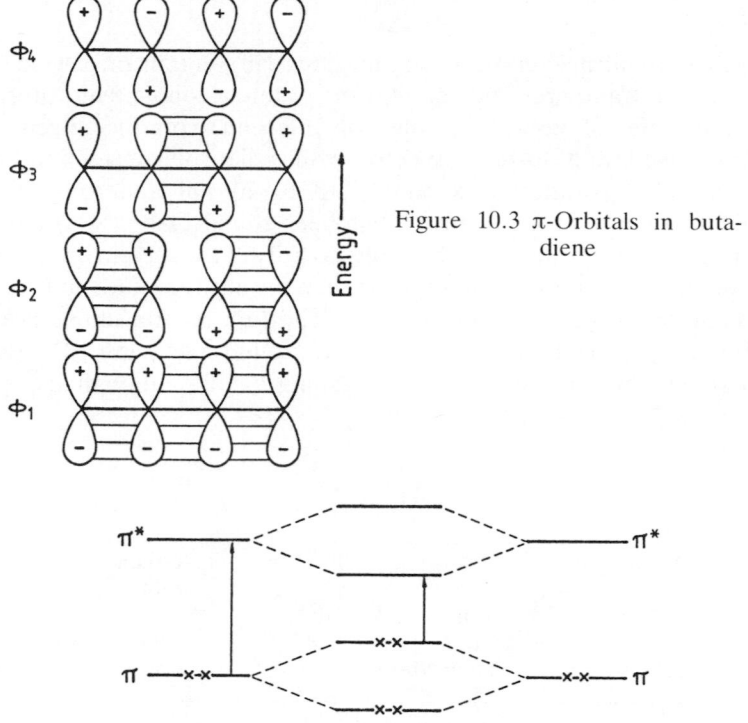

Φ_4

Φ_3

Φ_2

Φ_1

Energy

Figure 10.3 π-Orbitals in butadiene

Figure 10.4 The interaction of two π-orbitals in
butadiene

biphenyls are normally planar, with the π electron systems in the two rings conjugated to each other. If the two methyl groups occupy the o-positions on the same side of the molecule, steric hindrance twists one ring relative to the other and conjugation is no longer possible.

Carbonyl compounds

Aldehydes and ketones show absorption bands at about 290 nm; these are caused by a non-bonding electron on the oxygen atom being promoted to an antibonding π^* orbital. This (π^*, n) transition is shown in Figure 10.5.

Molecules in which the C=O bond is conjugated, such as $CH_2=CH$ $-CHO$, also show (π^*, n) transitions, but the conjugation stabilizes the π^* orbital, as in the dienes, and the transition is shifted to longer wavelengths.

Aldehydes and ketones also show bands attributed to (σ^*, n) and (π^*, π) transitions; these occur, as expected, at higher energies than the (π^*, n) transition. Similar transitions have been observed in comparable inorganic molecules and ions, such as nitrate and carbonate.

Figure 10.5 The (π^*, n) transition in H_2CO

Solvents

Most visible and ultraviolet spectra are measured in solution and a wide variety of solvents are transparent up to 200 nm; water, alcohols, chloroform and alkanes are commonly used. Non-polar solvents tend to produce spectra which are little changed from gas-phase spectra, while polar solvents tend to shift the wavelength of absorption maxima by up to 20 nm and to reduce the vibrational structure in the bands. It has been found that (π^*, n) bands are often shifted to shorter wavelengths as the polarity of the solvent increases; this is because the ground state is more polar than the excited state, and therefore interacts more strongly with the solvent. The (π^*, π) transitions are often shifted in the opposite direction, and here the ground state is less polar than the excited state. This can be of use in assigning bands, although there are a number of well-documented exceptions.

Table 10.2.

K^+, Ca^{2+}, $Sc^{3+} d^0$	Colourless	$Fe^{2+} d^6$	Green
Ti^{3+}	d^1 Pink	$Co^{2+} d^7$	Pink
V^{3+}	d^2 Blue	$Ni^{2+} d^8$	Green
Cr^{3+}	d^3 Green	$Cu^{2+} d^9$	Blue
Fe^{3+}	d^5 Yellow/brown	$Zn^{2+} d^{10}$	Colourless

C. Transition metal ions

It is well known that many transition metal ions are coloured. Table 10.2 gives the colours of some simple ions in the first transition series. In these ions the colour is caused by the incomplete d shell; the ions are coloured, except those with no $3d$ electrons (K^+, Ca^{2+}, Sc^{3+}) or with a complete $3d$ shell (Zn^{2+}).

The transition metal ions have five $3d$ orbitals, as shown in Figure 10.6; in the gas phase these orbitals are all degenerate, i.e. have identical energies. In solution the ions exist as complex ions and are surrounded by ligands; these liquids do not interact equally with all five $3d$ orbitals. Many complex ions, such as $Ti(H_2O)_6^{3+}$, have six ligands arranged in a regular octahedron round the central ion; the ligands interact more favourably with three of the d orbitals than with the other two, producing a splitting as shown in Figure 10.7. In $Ti(H_2O)_6^{3+}$ this splitting is about 20 400 cm^{-1}; the single $3d$ electron can be excited from one d orbital to another by light of about 490 nm, which lies in the visible region. In tetrahedral complexes the splitting pattern is reversed, as shown in Figure 10.8; two d orbitals are now more stable than the other three, but the energy difference still corresponds to light in the visible region. Excitation from one d orbital to another is always possible, provided that the d shell is neither empty nor completely filled.

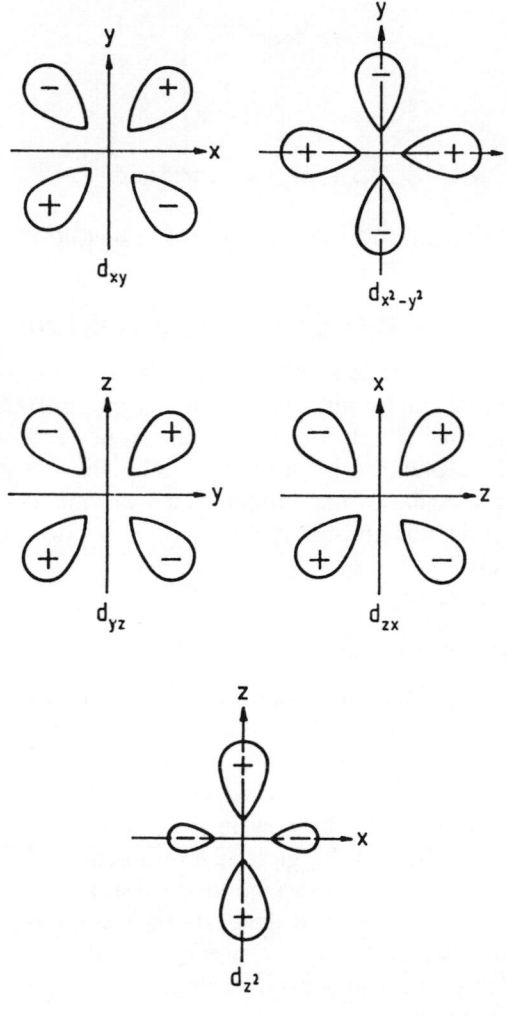

Figure 10.6 3d-orbitals

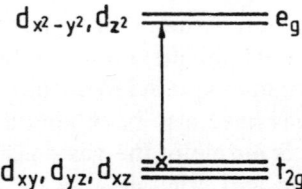

Figure 10.7 Absorption of light by $Ti(H_2O)_6^{3+}$

The excitation energies in transition metal complexes depend on the metal and on the ligand. Ligands may be arranged according to the splittings they produce; for some common examples the order is

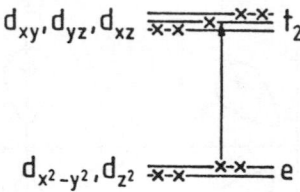

Figure 10.8 Absorption of light by $CuCl_4^{2-}$

$$I^- < Br^- < Cl^- < F^- < H_2O < NH_3 < NH_2CH_2CH_2NH_2 < CN^-,$$

where I^- produces the smallest splittings. Although the interaction of the ligands with the metal can be imagined as purely electrostatic, with negative ligands repelling the d electrons to different extents, this model has not proved capable of predicting all the observed splittings quantitatively. A molecular orbital model, in which some degree of covalency is allowed between ligands and the metal ion, has been found to explain the data more accurately.

D. Charge transfer transitions

Although many transition metal compounds are coloured because of d–d transitions between orbitals split by the ligand field, not all examples can be explained in this way. The ions VO_4^{3-}, CrO_4^{2-} and MnO_4^- are all coloured, even though their oxidation numbers correspond to their having no d electrons at all. In these cases a different type of excitation gives rise to the absorption of light, a charge transfer transition. Here an electron in a ligand orbital is excited to an unoccupied d orbital on the metal ion. The excitation energy of this transition depends on the energy of the vacant d orbital; for the series VO_4^{3-}, CrO_4^{2-}, MnO_4^- the transition energy drops steadily as the d orbital becomes more stable, i.e. as the ion becomes more oxidizing.

Charge transfer transitions have been observed in many transition metal complexes, including those which are coloured due to d–d transitions. The charge transfer bands are generally found at higher frequencies than the d–d bands, so it is common for transition metal ions to show a rather weak band in the visible region and a more intense band in the ultraviolet.

Charge transfer transitions have also been identified in a series of weakly bonded complexes. Iodine is purple in the gas phase and in solvents such as CCl_4. However, in benzene and ether it appears brown; it forms a weak complex with the solvent and the colour is due to a transition in which an electron in the solvent molecule is promoted to an antibonding orbital in the iodine molecule. The transition energies have been correlated with the ionization energies of the solvent molecules, and it is found that the lower the ionization energy of the solvent the longer the wavelength of the charge transfer band. This suggests that the electron is being transferred from the solvent to the iodine, and not the other way round.

E. Intensities

So far we have described absorption bands only by the wavelength at which they occur; in practice the intensities of the bands can also be of great value in assigning spectra. The measurement of spectra in the visible and ultraviolet regions is a very sensitive technique, and it is often possible to observe extremely weak bands corresponding to transitions which are formally forbidden. The intensities observed can vary considerably and may be useful in correlating bands in similar molecules.

As we saw in Chapter 8, the inherent probability of a transition is related to an integral R_{ij}, which measures the shift in charge during a transition

$$R_{ij} = \int \Psi_i^* R \Psi_j \, d\tau$$

where Ψ_i and Ψ_j are the final and initial wavefunctions. The molar extinction coefficient ε is proportional to the square of this integral. In condensed phases absorption bands are often very broad and the molar extinction coefficient must be summed over a range of frequencies.

In practice wavefunctions of the ground and exctied states of a molecule may not be well enough known for an exact prediction of the intensity of a band, but it is sometimes possible to show that the integral R_{ij} is zero for reasons of symmetry and that the transition is therefore forbidden. Whether a transition is forbidden is most readily calculated using group theory; the details are beyond the scope of this book, but the results in some common cases are given below.

The most important selection rule governing electronic transitions is concerned with electronic spin. In light elements spin–orbit coupling is very small and we can say that the quantum number for electron spin, S is well defined. Transitions in which this quantum number changes are forbidden, and so the selection rule is

$$\Delta S = 0.$$

This rule holds well in organic molecules, and as most stable organic molecules have ground states in which all the electron spins are paired, the only transitions which are observed are to excited states for which $S = 0$. These are singlet–singlet transitions. We shall see later that this rule is broken in the observation of phosphorescence. In heavy transition metal ions the rule holds less well and weak bands in which S changes can be observed.

Other transitions may be forbidden because of the symmetries of the initial and final electronic wavefunctions. In practice these transitions are observed, however, albeit with low intensities. This is because the symmetry of a molecule may be temporarily disturbed by a vibrational motion, and an integral which is zero when the atoms are in their equilibrium positions may no longer be so when the molecular shape is altered. These transitions are said to be vibronically allowed.

Charge transfer transitions involve a large shift in electronic charge, and these are allowed transitions which give rise to bands of high intensity. However d–d transitions in transition metal ions are forbidden both in octahedral and tetrahedral symmetries; they are observed only because of the vibrational motion in the ligands. They therefore give rise to low intensity bands; in this way intensities help to distinguish d–d and charge transfer bands in transition metal ion spectra.

In organic molecules the (π^*, n) transition in symmetrical aldehydes and ketones is forbidden, and is observed only very weakly. The transition is not formally forbidden in unsymmetrical cases such as CH_3CHO, although, because the $-CH_3$ and $-H$ groups affect the $C=O$ group very little, the observed intensity is still very low. The (π^*, π) transition in alkenes is allowed, however, and is found to be much more intense. This intensity is further increased if the double bond is conjugated.

So far we have considered only the absorption of light by molecules in the liquid phase, where they are free to move, and so adopt any angle to the incoming light. However, light can also be absorbed by solid samples, in which molecules are ordered in a regular pattern. If polarized light, whose electric field oscillates in only one plane, is now employed, then the variation of absorption with angle may now be studied; this can be of value in identifying transitions as some transitions are excited only by light polarized in a particular direction. For example, the (π^*, π) transition in ethene is excited only by light oscillating in the plane of the molecule, parallel to the $C=C$ bond. Once again group theory is needed to analyse each case fully.

F. Fluorescence and phosphorescence

When molecules absorb light in the visible and ultraviolet regions, excited states are formed. These do not usually have very long lifetimes, and in many cases the excitation energy is dissipated in the course of collisions into the vibrational and rotational energy of the whole sample. In this way thermal equilibrium is restored; the molecule has undergone non-radiative transitions.

If energy can only be dissipated slowly by collisions, however, then an excited state may also lose energy by emitting light; this gives rise to fluorescence and phosphorescence. Fluorescence occurs rapidly after the inital absorption of light, in times typically from 10^{-9} to 10^{-4} s, while phosphorescence involves a longer delay, from 10^{-4} to 10^2 s. As the non-radiative transitions with which emission competes depend on collisions, their rates are temperature-dependent; they become slower on cooling, so fluorescence and phosphorescence spectra are often measured below room temperature. The samples are usually solutions or solids.

When a molecule absorbs light, transitions are observed from the lowest vibrational level of the ground state to a series of vibrational levels of an excited

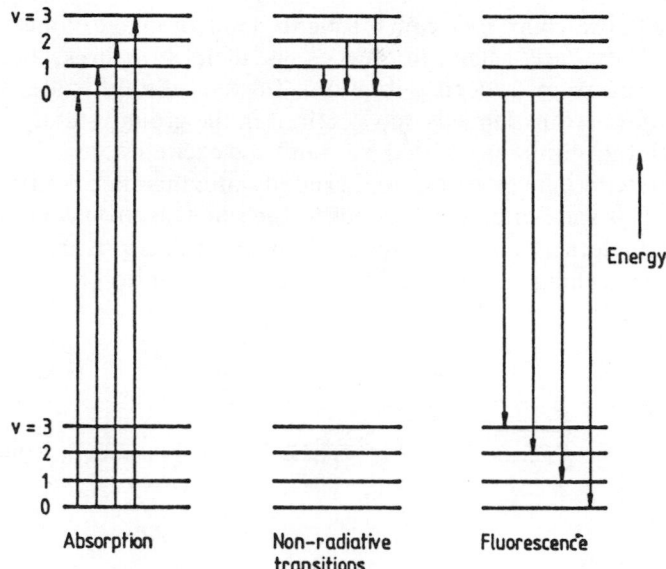

Figure 10.9 Fluorescence

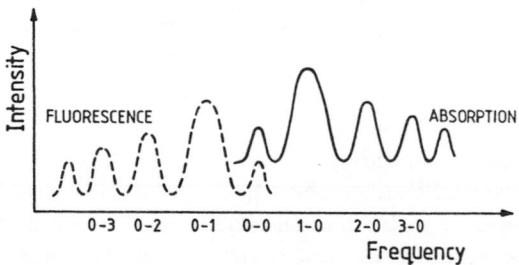

Figure 10.10 Fluorescence and absorption spectra

state (Figure 10.9). For most molecules at room temperature higher vibrational levels of the ground state are not significantly populated. If the ground state is a singlet ($S = 0$), the excited state will also be a singlet. Following excitation, the most rapid process is usually the dissipation of vibrational energy in the excited state through collisions. This produces excited state molecules in their lowest vibrational level. Dissipation of the electronic energy by collisions may be much slower, however, and spontaneous emission of light can occur. This is fluorescence. Transitions occur to a number of vibrational levels in the ground state, following the Franck–Condon principle. The fluorescence spectrum is shown in Figure 10.10, and is compared with the corresponding absorption spectrum.

Fluorescence always occurs at lower energies than absorption; this is familiar from fluorescent dyes, which absorb ultraviolet light and emit in the visible

136

region. The fluorescence spectrum is roughly a mirror image of the absorption spectrum. There are small differences in their structures, because the absorption spectrum reflects vibrations in the excited state while the fluorescence spectrum depends on vibrations in the ground state. The solvent may also interact differently with the ground and excited states.

Phosphorescence involves a second excited state; this state is a triplet, i.e. it has $S = 1$. This state often corresponds to the same electron configuration as the excited singlet state, but the spins of its unpaired electrons are now parallel, rather than cancelling each other. The triplet state is of lower energy than the singlet.

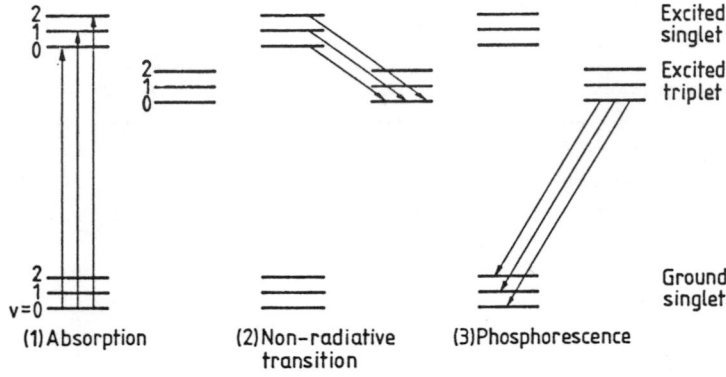

Figure 10.11 Phosphorescence

In phosphorescence non-radiative transitions cause the molecule to switch from the singlet excited state to the triplet state. This is called intersystem crossing and is formally forbidden. In practice it does occur to some extent, particularly if paramagnetic substances are present. The vibrational energy of the triplet state is now dissipated, just as in fluorescence (see Figure 10.11). If collisions cannot remove the electronic energy, the only fate of the molecule is to emit light and fall back to the ground state. This triplet-singlet transition is of course forbidden, if spin–orbit coupling is neglected; in practice it occurs very slowly and gives rise to the phosphorescence spectrum. Once again the spectrum shows structures due to vibrations in the ground state; more importantly, it gives information on the relative energies of the singlet and triplet states, which can be difficult to obtain in other ways.

G.   Summary

Molecules in condensed phases have electronic spectra which consist of broad bands, not individual lines. Sometimes they show vibrational structure, although often even this is lost. Measurement of ultraviolet spectra is rapid and convenient, and has been widely used for the identification of characteristic groups and for investigating molecular environments and interactions.

Organic molecules give important bands when multiple bonds are present; these bands become more intense, and of lower energy, on conjugation. Charge transfer transitions have been studied in many inorganic systems; d–d bands are important in transition metal complexes. The latter are formally forbidden, but in practice are observed with very low intensity.

Molecules sometimes re-emit light after absorption; fluorescence involves emission from the lowest vibrational state of a singlet excited state, and phosphorescence from the lowest vibrational level of a triplet excited state. Fluorescence occurs very rapidly, but phosphorescence, which is formally forbidden, can involve a longer delay.

H.   Problems

1. The ion $Cr(H_2O)_6^{3+}$ has an excited state 13 900 cm^{-1} above the ground state; what wavelength light would it absorb?

2. The ultraviolet absorption spectra of o- and p-substituted benzoic acids have been measured; in many cases it is found that the absorption in the o-isomers is at lower wavelengths than in the p-isomers. Why might this be?

3. The ions $OsCl_6^{2-}$, $OsBr_6^{2-}$ and OsI_6^{2-} show charge transfer bands at 27 000 cm^{-1}, 21 000 cm^{-1} and 15 000^{-1} respectively; they also show further bands at 45 000 cm^{-1}, 38 000 cm^{-1} and 31 000 cm^{-1} respectively. Comment on these values. The ion $PtBr_6^{2-}$ shows a band similar to the band in $OsBr_6^{2-}$ at 38 000 cm^{-1}, but no band similar to the one at 21 000 cm^{-1}. Why not?

4. How would one expect the electronic spectrum of liquid oxygen to differ from that of gaseous O_2?

5. How could an 'added brightness' be given to washing by using a detergent which makes clothes 'whiter than white'? Does this latter phrase have any meaning?

Chapter 11

Magnetic resonance spectroscopy

A. Introduction

It has long been known from atomic spectroscopy that electrons spin on their own axes and behave as tiny magnets; the splitting of the two D lines in the sodium spectrum is a familiar consequence of electron spin. It is also found that when the lines in the spectra of some atoms are studied at very high resolution, they are found not to be single, but to consist of several lines very close together. This hyperfine structure is caused by the fact that the nuclei of these atoms also have spin angular momentum, and hence magnetic moments.

In this chapter we shall be concerned with two techniques in which these small magnetic moments interact with an external magnetic field, nuclear magnetic resonance (NMR) and electron spin resonance (ESR). The spectroscopic transitions which are observed correspond to the realignment of the magnetic moment in a strong magnetic field.

For an NMR spectrum to be obtained, it is necessary that a molecule should contain at least one nucleus which has a magnetic moment; as we shall see below, not all nuclei have magnetic moments, but most common elements have at least one isotope which can be used to obtain NMR signals. NMR has therefore been applied to a very wide variety of chemical systems.

Electron spin resonance is more restricted in its application, as all closed-shell molecules have only paired electrons and therefore no resultant electronic spin. ESR spectra can only be obtained from species with unpaired electrons; examples include organic and inorganic free radicals, and transition metal ions. They do not, however, need to be stable indefinitely at room temperature, and ESR has been used to study reactive intermediates.

The actual magnitudes of the energies of the transitions observed in magnetic resonance experiments depend on the strength of the applied magnetic field; with the field strengths normally used in the laboratory, NMR signals are observed in the radio-frequency region of the spectrum. ESR signals occur at higher energies, as the magnetic moments of electrons are greater than those of nuclei, and are observed in the microwave region.

If spin resonance experiments could detect only the presence of certain nuclei or of unpaired electrons in molecules, then they would be of some analytical interest, but no great importance. The real significance of spin resonance lies in the fact that nuclei and unpaired electrons also experience small magnetic fields from the motions of other nuclei and electrons, within their own molecules and sometimes in adjacent molecules. These fields modify the appearance of the spin resonance spectrum, which can then be used to obtain a very clear picture of the precise chemical environment in a molecule. This detailed information has made NMR spectroscopy one of the most important and fruitful areas of research over the last 25 years, which has found applications in almost all branches of science.

B. Basic principles of NMR

As we have seen, the necessary condition for a nucleus to give an NMR signal is that it should possess spin angular momentum, and hence have a magnetic moment. Each nucleus has a definite spin angular momentum whose value is determined experimentally. Table 11.1 shows the values for some common nuclei. As is usual in quantum mechanics, angular momentum is quantized in units of $h/2\pi$; the angular momentum is described by a quantum number I such that

$$\text{Angular momentum} = \sqrt{I(I + 1)}\,\frac{h}{2\pi}$$

where I is an integer or half-integer.

Although there is no simple theory which allows us to predict the value of I for every nucleus, some general rules can be formulated:

(a) Nuclei where the sum of the number of protons plus neutrons is odd have half-integral spin.
(b) Nuclei where the sum of the number of protons plus neutrons is even have integral spin.
(c) Nuclei with an even number of protons and an even number of neutrons have zero spin.

These rules are explained in part by the fact that both the proton and the neutron have spin angular momentum, with $I = \frac{1}{2}$. For example, the ^{13}C nucleus contains six protons and seven neutrons, and has $I = \frac{1}{2}$. The ^2H nucleus contains one proton and one neutron, and has $I = 1$; the ^{32}S nucleus contains sixteen protons and sixteen neutrons, and has $I = 0$. In particular we should note that both the ^{12}C and ^{16}O nuclei are non-magnetic; this is of great importance in simplifying the ^1H-NMR spectra of organic compounds.

Nuclei with $I = 0$ have zero magnetic moments, but there is no simple way of predicting the magnetic moments of nuclei which do have spin angular momentum. The unit in which nuclear magnetic moments are usually

Table 11.1   Nuclear spins of some common nuclei

Isotope	Per cent natural abundance	Spin I
^{1}H	99.984	½
^{2}H	0.016	1
^{3}H	0	½
^{13}C	1.108	½
^{14}N	99.635	1
^{17}O	0.037	5⁄2
^{19}F	100	½
^{31}P	100	½
^{33}S	0.74	3⁄2

measured is the nuclear magneton μ_N, which is equal to $eh/4_\pi M_p$, where M_p is the mass of a proton; it has the value 5.05×10^{-27} J T^{-1} ($1JT^{-1} \equiv 1$ m^{2}A). The relationship between the angular momentum p and the magnetic moment μ is then written as

$$\mu = 2\pi \frac{g\ \mu_N\ p}{h},$$

where g is a numerical factor whose value has to be determined experimentally for each nucleus. g values are usually, though not invariably, positive. The ratio of the spin angular momentum to the magnetic moment is called the magnetogyric ratio, γ:

$$\gamma = 2\pi \frac{g\ \mu_N}{h}.$$

We have seen that the angular momentum p is given by the quantum number I; therefore

$$\mu = g\mu_N \sqrt{I(I+1)}.$$

When a nucleus is placed in a magnetic field, it experiences a torque which causes it to precess about the field direction, just as a gyroscope precesses in a gravitational field (Figure 11.1). The frequency of this precession depends on the applied field and is called the Larmor frequency. The angle at which the nucleus precesses may not take any value, but, as is usual in quantum mechanics, is such that the component of the angular momentum in the direction of the field is $(h/2\pi)M_I$, where M_I is another quantum number and may take the values I, $(I-1)$, ..., $(1-I)$, $-I$. Thus for a proton, $I = ½$, and so M_I may take the values $+½$ and $-½$, which we may speak of roughly as the nucleus being aligned with or against the field. For a deuterium nucleus, $I = 1$, and so three orientations of the nucleus are possible, with M_I values of $+1$, 0 and -1.

The potential energy of a nuclear magnetic moment μ in an external field B is given by

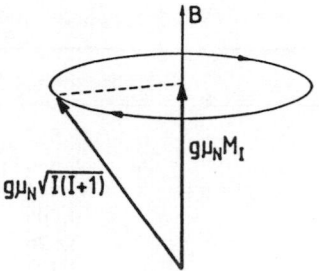

Figure 11.1 Precession of the
nuclear magnet in a magnetic
field

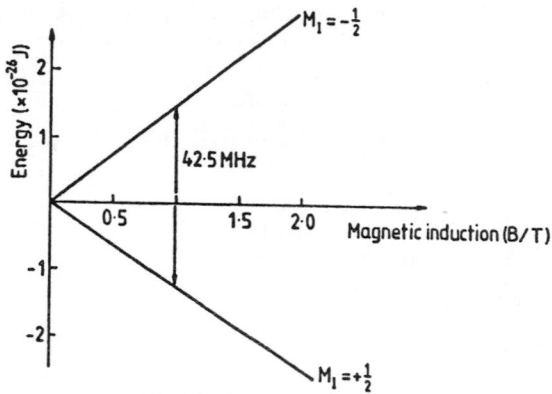

Figure 11.2 Energies of the two states of the proton
in a magnetic field

$$E = - B \mu \cos \theta,$$

where θ is the precession angle. (Strictly B is the magnetic flux density; if B is measured in teslas then E is in joules.) The quantity $\mu \cos \theta$ is simply the component of the magnetic moment in the field direction, which is the component of the angular momentum in the field direction multiplied by the magnetogyric ratio. Therefore,

$$E = - B h M_I \gamma$$
$$= - B M_I g \mu_N.$$

Figure 11.2 shows the energies of the two states of the ^1H nucleus in a magnetic field. Transitions between these two states are magnetic-dipole-allowed, and the transition energy is

$$\Delta E = B g \mu_N [½ - (-½)]$$
$$= B g \mu_N.$$

This is the energy required to reverse the direction of the nuclear spin. The

Table 11.2   NMR frequencies at 1T

Nucleus	NMR frequency (MHz)
^{1}H	42.557
^{2}H	6.536
^{3}H	45.414
^{13}C	10.705
^{14}N	3.076
^{17}O	5.772
^{19}F	40.055
^{31}P	17.235
^{33}S	3.266

frequency of the radiation corresponding to this energy is

$$\nu = \frac{B \, g \, \mu_N}{h},$$

which is equal to the Larmor precession frequency. For a proton in a field of 1 T, the frequency is

$$\nu = \frac{1 \times 5.585 \times (5.05 \times 10^{-27})}{(6.63 \times 10^{-34})} \text{ Hz}$$

$$= 4.25 \times 10^7 \text{ Hz}$$

$$= 42.5 \text{ MHz},$$

which is in the radiofrequency region of the spectrum.

Similar calculations can be carried out for other nuclei; when $I > \frac{1}{2}$, the selection rule for transitions is $\Delta M_I = \pm 1$. As each nucleus has its own g value, so it has its own resonant frequency at any given field strength. Typical values for some common nuclei are given in Table 11.2; as values of g vary quite widely, so do the resonant frequencies, and any NMR experiment detects the signals from one type of nucleus only.

One of the practical problems of NMR is that it is inherently an insensitive technique. When a sample is irradiated at the resonant frequency, transitions are induced in which nuclei parallel to the field become antiparallel, and vice versa. These two processes, stimulated absorption and stimulated emission, are of equal probability, and a net absorption of energy is detected only because one orientation has a greater equilibrium population than the other. The energy differences in NMR are so small (compared with kT) that the numbers of nuclei in each orientation are almost exactly equal. For example, for a ^{1}H nucleus with a resonant frequency of 60 MHz, the Boltzmann distribution law gives

$$n_-/n_+ = \exp \frac{-h\nu}{kT}$$

$$= \exp - \frac{(6.63 \times 10^{-34}) \times (60 \times 10^6)}{(1.38 \times 10^{-23}) \times (300)}$$

$$= \exp -(1 \times 10^{-5})$$

$$= 0.99999.$$

The observation of an NMR signal depends on this slight excess of 1 part in 10^5 of one spin state over the other; but for most purposes in the interpretation of spectra, it is clearly safe to assume that all nuclear spin states are equally populated.

The intrinsic probability of an NMR transition is independent of the chemical environment of a nucleus, and the intensity of an NMR line is, under optimum conditions, proportional to the number of nuclei which give rise to the transition. NMR may thus be used in quantitative analysis. For example, by comparing the signal intensity with that of a standard, the water content of silica balls, used as catalysts, may be determined from their proton absorption. Alternatively, the relative intensities of lines in a spectrum can be measured; in this way the aliphatic to aromatic ratios in oil mixtures can be determined, because aliphatic and aromatic hydrogen nuclei resonate, as we shall see below, at slightly different frequencies.

In NMR the sample is usually in the liquid state, either pure or in solution. It is also possible to use solid samples, although as we shall see later, the appearance of the spectra of solids is markedly different from that of liquids.

The first nucleus to be used widely in NMR experiments was ^1H, the proton; this isotope has a high natural abundance, and hydrogen is found in a wide variety of compounds. Later other nuclei such as ^{19}F, ^{23}Na and ^{31}P were employed, and as experimental techniques improved it became possible to use isotopes such as ^{13}C, where the natural abundance is only 1 per cent.

C. Chemical shifts

In practice it is found that all the nuclei of any given isotope do not resonate at exactly the same frequency; the resonant frequency is affected slightly by the chemical environment of the nucleus. Figure 11.3 shows the proton NMR spectrum of $HCOOCH_3$; the spectrum is plotted with the frequency held constant and the magnetic field strength being varied to bring all the nuclei to resonance in turn. The spectrum consists of two peaks, corresponding to the $-CH_3$ and $-\underset{\underset{O}{\|}}{C}-H$ protons; the intensities of the two peaks are in the ratio of 3:1, reflecting the relative numbers of each sort of proton.

The splitting of the NMR spectrum into two peaks is caused by the fact that each nucleus is partially shielded from the external field by the electrons which surround it. The external field causes these electrons to move and their motion sets up a new field, which by Lenz's law opposes the original field. The magnitude of this shielding depends on the local electron density, and so nuclei

144

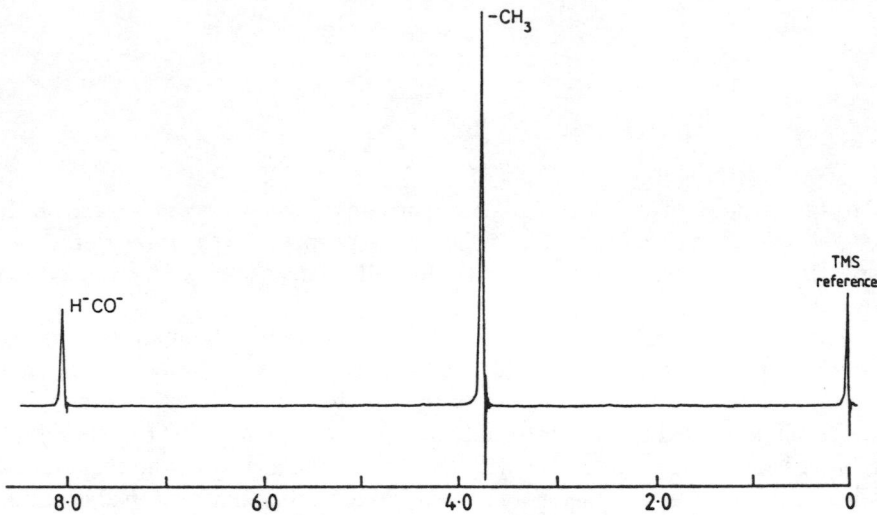

Figure 11.3 NMR spectrum of HCOOCH₃ with the δ sale in ppm

in different environments are shielded to different extents, i.e. have different 'chemical shifts'.

As chemical shifts are very small in comparison with the fields used in spectrometers, it is normal to measure the position of a peak relative to that of an arbitrarily chosen standard. Tetramethylsilane (TMS), $(CH_3)_4Si$, is added to samples to act as a standard for both ^1H and ^{13}C spectra, as it is very unreactive and gives a single peak which does not normally overlap with other peaks.

Chemical shifts are proportional to the field at which a spectrometer operates. There is no standard field strength at which all spectrometers operate and so chemical shifts are measured not in field or frequency units, which would be machine-dependent, but as a dimensionless ratio. The chemical shift of a nucleus i, δ_i, is given by

$$\delta_i = \frac{B_{TMS} - B_i}{B_{TMS}},$$

where B_i and B_{TMS} are the fields required to bring i and TMS to resonance. To bring the numbers to a convenient range, the δ values are then multiplied by 10^6, and expressed as parts per million (p.p.m.). The range of chemical shifts found for ^1H nuclei covers about 15 p.p.m.; the ^{13}C nucleus has a higher electron density and its range of chemical shifts is about 300 p.p.m.

Chemical shifts for H nuclei are also sometimes expressed on the τ scale; here TMS is assigned a shift of 10 p.p.m. and peaks at lower fields have lower τ values. The τ and δ scales are related:

$$\delta = 10 - \tau,$$

as shown in Figure 11.4, the spectrum of CH_3COOH. Notice that the proton attached to the electronegative oxygen atom has a relatively low electron

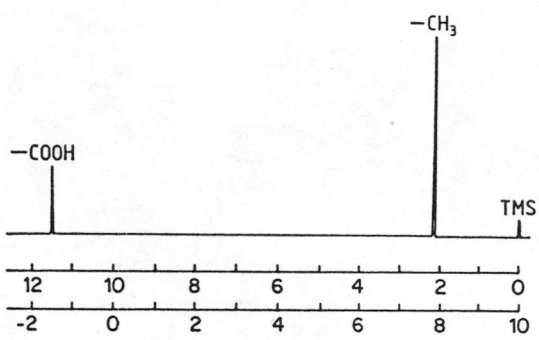

Figure 11.4 ^{1}H-spectrum of CH_3COOH (in $CDCl_3$)

Table 11.3 Chemical shifts for CH_3X

X	τ
$-Si(CH_3)_3$	10
–aliphatic group R	9.1
$-CR=CR_2$	8.3
–COR	7.9
–I	7.84
$-NR_2$	7.8
–aromatic hydrocarbon	7.7
–Cl	6.95
$-N^+R_3$	6.7
–OCOR	6.3
–F	5.74
$-NO_2$	5.72

density, and is therefore not shielded as much as the $-CH_3$ protons, with a higher electron density.

Chemical shifts are often the most useful features of NMR spectra. A very wide variety of organic compounds have been studied by NMR spectroscopy; at first ^{1}H-NMR was used exclusively, but more recently ^{13}C spectra have become more routinely available. It is found that many functional groups have a characteristic chemical shift, whose value is affected relatively little by substituents; some examples are given in Table 11.3. The use of such tables has made NMR a particularly powerful technique in the analysis of unknown compounds; it can often reveal structural features which are difficult to detect by other methods.

The chemical shifts of many functional groups may be rationalized by simple electronegativity arguments; thus the introduction of a fluorine atom into a group decreases the electron density, producing less shielding and a shift to lower field. Deviations from these simple predictions are often found in aromatic systems, where the external magnetic field causes the mobile π electrons to circulate. These 'aromatic ring currents' are shown in Figure 11.5;

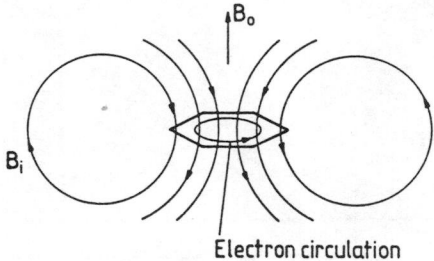

Electron circulation

Figure 11.5 Aromatic ring currents

they cause protons attached directly to the benzene ring to resonate at low fields, as the field from the ring reinforces the external field outside the ring. However, protons immediately above or below the ring would resonate at high fields, as here the field from the ring currents opposes the external field. This effect has been observed in the cyclophanes,

$$CH_2 -C_6H_4 -CH_2$$
$$CH_2-(CH_2)_n-CH_2$$

where the CH_2 protons show a wide variety of chemical shifts, depending on their position relative to the benzene ring. Ring current shifts are often large and have been used to obtain information on the conformations of biochemical molecules; the resonances of some groups are found at unusually high fields, which can be interpreted in terms of their proximity to an aromatic ring in some other part of the molecule.

Chemical shifts are also strongly affected by hydrogen bonding. Hydrogen bonding shifts the proton resonance to low field; the effect can be large, of the order of 10 p.p.m. The hydrogen bonding may be either intramolecular, as in nitrophenol, or intermolecular. Chemical shifts in hydrogen-bonded systems are often strongly concentration-dependent; this dependence can give useful information on the hydrogen-bonding, but means that lines from $-OH$ and $-NH-$ protons have to be interpreted with great care in structure determination.

D. Spin–spin coupling

Many NMR spectra contain more lines than can be explained on grounds of chemical shift differences alone; this further structure arises from the interactions between the magnetic nuclei within a molecule. Figure 11.6 shows the ^1H-NMR spectrum of $CHCl_2-CH_2Cl$; absorption occurs in two different parts of the spectrum, corresponding to the two hydrogen environments, but each of these absorptions is split into a series of well-resolved lines.

There is a direct interaction between any pair of magnetic dipoles in a molecule, but in liquid samples the rapid molecular tumbling which occurs

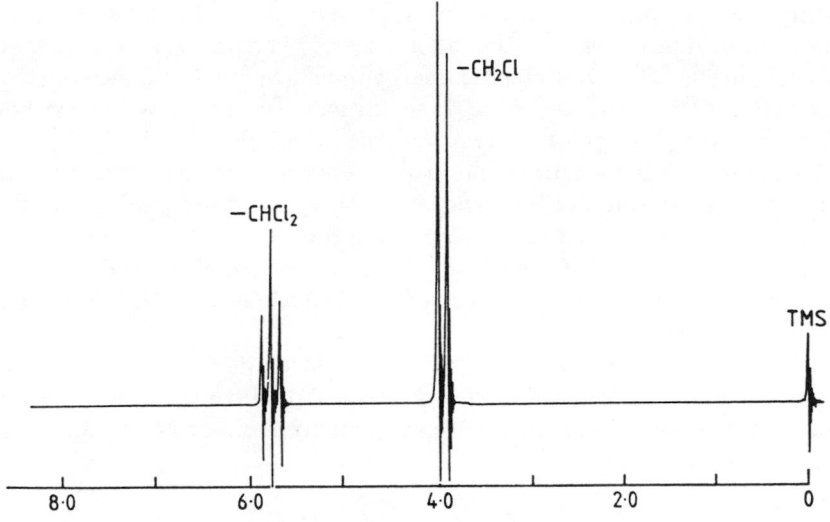

Figure 11.6 ^{1}H Spectrum of $CHCl_2$–CH_2Cl on the delta scale

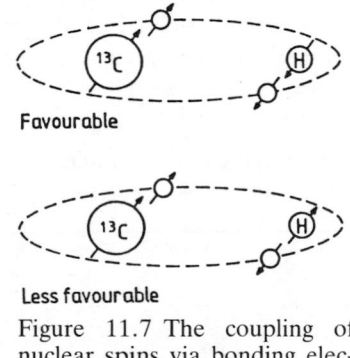

Figure 11.7 The coupling of nuclear spins via bonding electrons

averages this interaction to zero. However, two magnetic nuclei can also interact via the bonding electrons between them; this interaction is not averaged to zero by molecular tumbling and gives rise to the fine structure visible in Figure 11.6. We can illustrate this mechanism by considering a ^{13}C and a ^{1}H nucleus joined by a single covalent bond (Figure 11.7). The bond consists of two electrons whose spins (and therefore magnetic moments) are opposed—by the Pauli exclusion principle. Each nucleus attracts both electrons electrostatically, but there is a further magnetic interaction, which is favourable if the electron and nuclear magnetic moments are parallel to each other. If the ^{13}C and ^{1}H nuclei have opposed spins, then one electron will move fractionally closer to the ^{13}C nucleus and the other to the ^{1}H nucleus, so that both magnetic interactions are favourable. This cannot occur if the ^{13}C and ^{1}H nuclei are parallel; if one electron has a favourable interaction with one nucleus, then the other electron at the opposite end of the bond must have an unfavourable interaction with the other nucleus. The parallel arrangement of nuclear spins therefore has a higher energy than the paired arrangement.

The effect of this coupling on the ^1H spectrum is to produce two lines of equal intensity, corresponding to the two possible orientations of the ^{13}C nucleus. The magnitude of the splitting is not dependent on the magnetic field strength and is usually measured in frequency units (hertz) and not parts per million. The splitting is called the spin–spin coupling constant, J. The coupling has an identical effect on the ^{13}C spectrum; two lines of equal intensity are seen, corresponding to the two orientations of the ^1H nucleus, and the splitting is the same as in the ^1H spectrum.

In fact, of course, the splitting of the energy levels does not necessarily split the peaks, but a more careful analysis does show how this arise. The spin energy of two nuclei, A and X, of spin quantum number $M_I(A)$ and $M_I(X)$ respectively is given by

$$\frac{E_{spin}}{h} = - (M_{I(A)}\nu_A + M_{I(X)}\nu_X) + M_{I(A)}M_{I(X)}J_{AX},$$

where J_{AX} is the spin–spin coupling constant between the nuclei. There are thus four possible values for the spin energy, depending on the various values of the spin quantum numbers. These four values are usually labelled as $\alpha\alpha$, $\alpha\beta$, $\beta\alpha$ and $\beta\beta$, as in Table 11.$\bar{4}$.

Table 11.4   Spin energies for combinations of nuclear spin

$M_{I(A)}$	$M_{I(X)}$	Label	E_{spin}/h
$+ \frac{1}{2}$	$+ \frac{1}{2}$	$\alpha\alpha$	$- \frac{1}{2}(\nu_A + \nu_X) + J_{AX}/4$
$+ \frac{1}{2}$	$- \frac{1}{2}$	$\alpha\beta$	$- \frac{1}{2}(\nu_A - \nu_X) - J_{AX}/4$
$- \frac{1}{2}$	$+ \frac{1}{2}$	$\beta\alpha$	$+ \frac{1}{2}(\nu_A - \nu_X) - J_{AX}/4$
$- \frac{1}{2}$	$- \frac{1}{2}$	$\beta\beta$	$+ \frac{1}{2}(\nu_A + \nu_X) + J_{AX}/4$

There are two transitions corresponding to absorption of energy by nucleus A, for which $\Delta M_{I(A)} = -1$, $\Delta M_{I(X)} = 0$. These are $\beta\alpha \to \alpha\alpha$, for which $\Delta E/h = \nu_A - \frac{1}{2}J_{AX}$ and $\beta\beta \to \alpha\beta$, for which $\Delta E/h = \nu_A + \frac{1}{2}J_{AX}$. If X did not exist, the peak would be at ν_A. Spin–spin coupling causes this peak to split into two, one peak $\frac{1}{2}J_{AX}$ above and the other $\frac{1}{2}J_{AX}$ below the original.

It is not necessary that the two nuclei should be directly bonded to each other; spin–spin coupling is seen between the ^1H nuclei in $CHCl_2-CHO$, as in Figure 11.8. Note that the splittings of the two pairs of lines are equal. Generally it is found that spin–spin coupling constants decrease as the internuclear distance increases; some typical values are given in Table 11.5. The sign convention is that J is positive where the antiparallel arrangement of nuclei is more stable and negative where it is less stable.

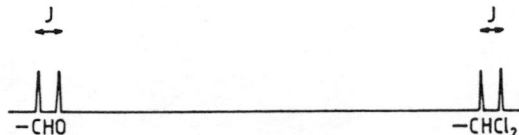

Figure 11.8 Spin–spin coupling in the proton
NMR spectrum of $CHCl_2CHO$

Table 11.5 Spin–spin coupling constants (H_3)

$^1H-^1H$	280
C^1H-C^1H	6–8
C^1H-C-C^1H	1–2
$C^1H-C-C-C^1H$	<0.5

Many molecules contain more than two magnetic nuclei which can spin-couple with each other; we can distinguish a number of situations, each with its own characteristic splitting pattern.

A particularly simple spectrum arises where two or more nuclei are in exactly equivalent chemical environments, such as the three protons in a $-CH_3$ group. Although the magnetic moments of the nuclei do interact, it can be shown that all transitions allowed by the selection rules occur at exactly the same energy, and so only one line is observed in the spectrum. This simplifies the appearance of many spectra and is summarized in the rule that spin-splitting is not observed between equivalent nuclei.

A more complex splitting pattern is produced when one proton couples to two other equivalent protons, as in $CH_2Cl-CHO$. If we consider the magnetic environment of the $-CHO$ proton, there are now three possibilities. Both the protons in the $-CH_2Cl$ group could be parallel to the field, both could be antiparallel, or one could be parallel.and the other antiparallel. The third possibility is twice as likely as the other two, as there are two ways in which it can be achieved. The absorption of the $-CHO$ proton is therefore split into three lines, with intensities in the ratio of 1:2:1. The $-CH_2Cl$ protons couple to the single $-CHO$ proton, and so give rise to just two lines, with a different chemical shift.

If a nucleus interacts with three equivalent spins, then four lines are obtained in the spectrum, with intensities in the ratio of 1:3:3:1.

In all the analyses above we have treated spin-splitting as a small perturbation superimposed on the general pattern of chemical shifts in a spectrum. This is accurate, provided that the chemical shift between the nuclei which couple is much greater than the spin-coupling constant. If the spin-coupling constant and the chemical shift are of comparable magnitude, then deviations occur, both in the positions of the lines and their intensities; this is illustrated in Figure 11.9. As the chemical shift increases with applied

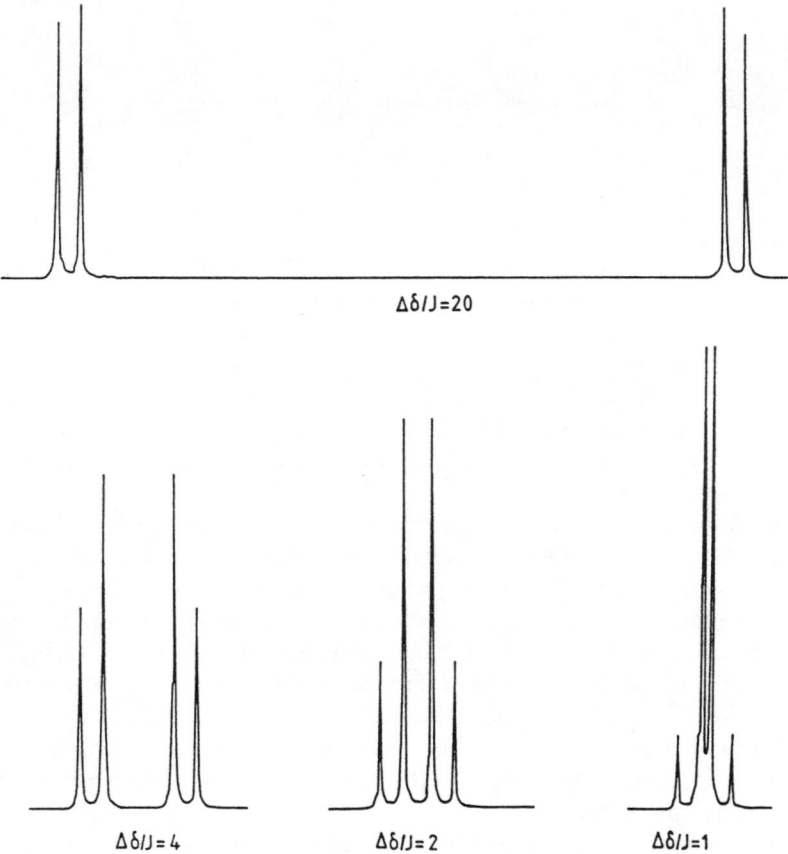

$\Delta\delta/J=20$

$\Delta\delta/J=4$ $\Delta\delta/J=2$ $\Delta\delta/J=1$

Figure 11.9 Computer-simulated spectra for the interaction of two protons

field, whereas the spin-coupling constant does not, these deviations are particularly marked when spectrometers operate at low magnetic fields. This is one reason why the present trend is to build spectrometers operating at very high fields.

The relative abundances of the ^{1}H and ^{13}C nuclei have an important effect on the spin coupling which is observed in the NMR spectrum of organic compounds. Virtually all naturally occurring hydrogen is ^{1}H, whereas ^{13}C represents only about 1 per cent. of all naturally occurring carbon, the rest being the non-magnetic ^{12}C. In consequence, all ^{1}H-NMR spectra show spin coupling between pairs of hydrogen atoms, but not between hydrogen and carbon atoms, as almost all ^{1}H nuclei are bonded to non-magnetic carbon atoms. Similarly, ^{13}C spectra show spin coupling between carbon and hydrogen atoms, as all carbon atoms are bonded to magnetic hydrogen atoms, but not between pairs of carbon atoms, as almost all ^{13}C nuclei are bonded to non-magnetic ^{12}C nuclei. Isotopic enrichment can, of course, be used to obtain carbon–carbon coupling constants.

The appearance of a spin multiplet can be used to give important information on the structure of an unknown molecule. The most obvious parameter is the number of lines in the multiplet and the intensity distribution, giving direct information on the number of interacting nuclei. Thus the appearance of a 1:3:3:1 quartet in a ^{13}C spectrum could indicate the presence of a $-CH_3$ group. In practice, this is not always as useful as it might be thought. Lines of low intensity can be difficult to distinguish from background noise, and intensity distributions do not necessarily follow the simple rules, so the appearance of a spin multiplet is not always conclusive evidence on its own.

The magnitudes of spin–spin coupling constants can be of value in structure determination. These constants have values which are often characteristic of the stereochemistry of a molecule: e.g. the J value for two protons in a $-CH=CH-$ group is about $+10$ Hz if the protons are cis, and about $+18$ Hz for the trans arrangement. Measurement of spin splitting therefore allows the stereochemistries of substituted alkenes to be determined. Coupling constants can also be used to give conformational information in more flexible systems; e.g. the value of J for two protons in a $>CH-CH<$ fragment depends strongly on the dihedral angle between the two C–H bonds (see Figure 11.10). Although the precise shape of the graph depends on the nature of the substituents, the method is accurate enough to distinguish conformations in saturated ring compounds.

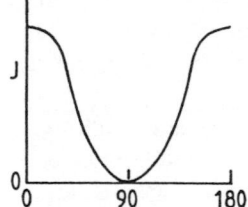

Figure 11.10 Variation of spin–spin coupling with torsion angle

E. Rate processes

One of the most remarkable features of NMR is the extreme narrowness of the lines; it is quite common to observe lines of width less than 1 Hz when the resonant frequency is of the order of 10^8 Hz. The uncertainty principle requires that, for a line to have a width less than 1 Hz, the energy state of the nucleus must remain constant for a time greater than of the order of 1 second. Many chemical processes occur faster than this, and they can modify the appearance of an NMR spectrum.

We can illustrate this effect by considering the molecule $(CH_3)_2N-N=O$, whose proton spectrum is shown in Figure 11.11. At room temperature rotation about the N–N bond is very slow and the two $-CH_3$ groups are in different chemical environments, with one cis and the other trans to the $-N=O$ bond. The spectrum consists of two sharp lines. The rate of rotation about the N–N bond increases with temperature, and as the average lifetime of the

152

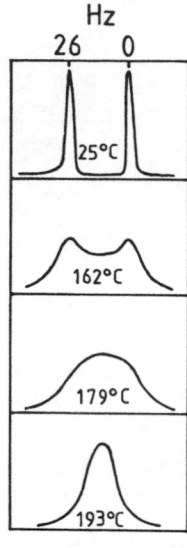

Figure 11.11 Effect of temperature on the ^{1}H spectrum of $(CH_3)_2N-N=O$

molecule in one energy state decreases, so the linewidth increases. As the temperature continues to rise, the line broadening becomes comparable with the chemical shift difference between the two $-CH_3$ groups and the peaks begin to merge. However, as the temperature rises further, a single peak is formed, whose width decreases with further temperature increases. Here the rate of rotation has become rapid compared with the chemical shift difference (expressed in hertz), and the chemical shift of this line is an average of the chemical shifts of the conformations of the $-CH_3$ groups.

Narrow lines are obtained in NMR spectra when a nucleus undergoes changes in its environment which are very slow, in which case each conformation gives rise to a line, or when it undergoes very rapid changes, when only a single line, representing its average environment, is seen. The key question here is what is meant by fast and slow. Processes with half-lives greater than 1 second do not normally give measurable broadening of lines and are therefore 'slow'. Similarly, chemical shift differences are normally less than 10^3 Hz and so processes with half-lives much less than 10^{-3} seconds are 'fast'. Processes with intermediate lifetimes, of the order of 10 ms, have a marked effect on NMR lineshapes and widths, and analyses of spectra have given detailed information on the kinetics of a variety of chemical reactions and restricted rotations.

Common examples of very rapid chemical exchanges include the interconversion of chair forms in cyclohexane rings and proton exchange in acidic solutions. In cyclohexane it is known that each hydrogen atom occupies either an equatorial or an axial position in the ring, but the rate at which one chair form flips into the other form, thereby exchanging axial and equatorial forms, is such that only one line is seen in the NMR spectrum at room temperature, at a position halfway between those expected for an equatorial and axial proton respectively. Similarly, the proton spectrum of CH_3COOH in water shows only two lines, one from the $-CH_3$ group and one from the $-COOH$ and H_2O protons. These exchange rapidly: $CH_3COOH + H_2O \rightleftharpoons CH_3COO^- + H_3O^+$, and it is not possible to observe separate lines for the $-COOH$ and H_2O environments. When a nucleus moves very rapidly between two non-equivalent positions, the value of the 'average' chemical shift can sometimes be used to obtain the equilibrium constant, if the chemical shifts of the positions separately are known.

A further consequence of rapid chemical exchange is the collapse of spin–spin coupling. In the proton spectrum of pure CH_3OH, the $-CH_3$ absorption is split into two lines by the neighbouring $-OH$ group. When mineral acid is added, the $-OH$ proton undergoes rapid exchange, which causes fast reorientation of its magnetic moment. The $-CH_3$ group now experiences only the 'average' orientation of the $-OH$ proton, and its signal collapses to a single line.

F. Double resonance

The effects of spin–spin coupling in a spectrum may also be removed by irradiating the sample with a strong radiofrequency field. This is one example of double resonance, in which two magnetic nuclei are simultaneously irradiated at their resonant frequencies.

In spin decoupling the sample is irradiated strongly with the resonant frequency of one of the magnetic nuclei, which causes this nucleus to undergo very rapid reorientation in the magnetic field, with the result that adjacent nuclei no longer experience spin coupling and their resonances collapse to a single line. This can be very desirable when a nucleus is coupled to several other nuclei, giving a complex pattern of lines which may overlap strongly with each other; analysis may only be possible when spin decoupling is used. The technique can also be used to establish which nuclei in a molecule couple together, and hence to obtain structural information. Spin decoupling can also give information on the relative signs of spin coupling constants; only their magnitudes can be deduced from simple splitting patterns.

A particularly important example of spin decoupling is the use of broad-band decoupling in ^{13}C-NMR. Here the sample is irradiated strongly over a range of frequencies covering all proton absorptions, and the ^{13}C spectrum is then observed normally. The effect is to remove all the proton fine structure from the spectrum; this structure is of little analytical value and the intensity of each nucleus is concentrated into a single line. The low natural abundance of ^{13}C means that no carbon–carbon spin coupling is observed, and so ^{13}C spectra are especially free of complexity.

Another type of double-resonance experiment is 'spin tickling'. It is similar to spin decoupling, but the irradiation used is much weaker. The weak irradiation of a single resonance line splits all transitions which share a common energy level with the line into doublets. Spin tickling can be of great value in the assignment of complex spectra; it can also be of help in determining the signs of J values.

G. NMR of solids

All the examples that we have considered so far have involved liquid samples, where there is rapid molecular tumbling. Thus tumbling removes from the spectra the effects of the direct dipolar interaction between magnetic nuclei, which is averaged by the motion to zero. However, in solid samples there is no molecular tumbling, and the direct interaction of magnetic dipoles now dominates the spectra, which consist of broad bands rather than a series of sharp lines.

The interaction energy of two magnetic dipoles separated by a distance r and at an angle θ to the line connecting them is given by

$$E = \frac{k(3 \cos^2\theta - 1)}{r^3}.$$

The value of k depends on the size of the dipoles and whether the nuclei are equivalent or not, but for two protons in the same molecule, the value of E may be of the order of 50 kHz, when expressed in frequency units. This interaction energy is far greater than those involved in chemical shifts and spin–spin coupling.

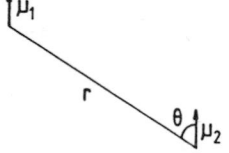

Figure 11.12 The interaction of magnetic dipoles

The NMR spectrum of a molecular solid containing two magnetic nuclei consists of two peaks, as the energy of one nucleus depends on the orientation of the other. The separation of the peaks is large and the peaks themselves are broad, because of unresolved interactions with more distant nuclei. The peak separation is a function of the angle θ, and so for a single crystal the NMR spectrum varies if the crystal is rotated in the magnetic field. In a poly-crystalline material, all possible values of θ are represented, and the resulting spectrum is an envelope. The width of the spectrum now depends only on the value of $1/r^3$; measurement of spectra is an important method of determining the hydrogen–hydrogen distances in solids, which are not easily measured by diffraction methods.

Similar results are obtained from crystals with more than two magnetic nuclei; the precise shape of the bands depends on the number and relative positions of the nuclei. Observation of band shapes can therefore be used to identify what species are present in a solid: e.g. NMR shows that solid hydrated HNO_3 contains H_3O^+ ions, with an equilateral triangular arrangement of hydrogen atoms, rather than separate H_2O and HNO_3 molecules.

Studies of linewidths of solid NMR spectra can also give information on molecular motion in solids. If NMR spectra are measured as a function of temperature, it is often found that the linewidth undergoes marked changes at temperatures well below the melting point. These changes can be attributed to the onset of some motion within the crystal lattice; e.g. the spectrum of benzene narrows appreciably at around $-170\ °C$, and this is interpreted as showing that above this temperature the molecules are free to rotate within the crystal lattice.

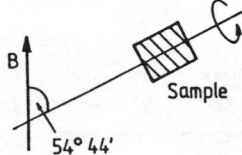

Figure 11.13 Magic angle spinning

Dipolar broadening so dominates the NMR of solids that no details of chemical shifts can be seen. These details are precisely what makes liquid state NMR such a sensitive tool for determining molecular structure, and recently considerable efforts have gone towards measuring solid state spectra with the dipolar broadening removed. This can be achieved by 'magic-angle spinning', in which the sample is spun very rapidly at an angle of $54°\ 44'$ to the magnetic field (Figure 11.13). As we saw above, the interaction between the two magnetic dipoles is proportional to the term $(3\cos^2\theta - 1)$; when θ is $54°\ 44'$, the 'magic angle', this term becomes zero, and so the dipolar broadening vanishes. For magic-angle spinning to be successful, the sample must be spun at several kilohertz—fast enough for the line between the nuclei to experience only its 'average' inclination to the field. This presents considerable technical difficulties. Nevertheless, the solid state spectra have been obtained for a variety of solids, and they show linewidths narrow enough to allow chemical shifts to be distinguished. Interpretation of the spectra is not always easy, as the chemical environment now includes interactions from adjacent molecules, and not just a single molecule, and the intensities are no longer quantitatively reliable, but it seems probable that high resolution solid state NMR will become a very important technique in the future.

H. Relaxation times

The equilibrium populations of the various energy levels encountered in spectroscopy are given by the Boltzmann distribution law, where the population decreases exponentially as the energy increases. In most forms of spectroscopy it is safe to assume that if these populations are temporarily disturbed in some way, e.g. by the absorption of light, then equilibrium will be restored fairly rapidly. In NMR this is frequently not the case, and after the nuclei in a sample are reorientated by radiofrequency field, it can take many seconds for the equilibrium populations to be restored. The processes by which the magnetization returns to equilibrium, or 'relaxes', have been carefully studied, partly because they have important experimental consequences, and also because they can give information on molecular motions.

For relaxation to take place, i.e. for the nuclei to reorientate themselves, they must experience a suitable magnetic field. Several different relaxation mechanisms have been identified: the motions of neighbouring magnetic nuclei can cause relaxation, and this is often the dominant mechanism in organic molecules. Paramagnetic species also cause very efficient relaxation, and organic radicals and transition metal ions are sometimes used for this purpose. Nuclei with spin $I > \frac{1}{2}$ have quadrupole moments, and these can induce relaxation. In heavy atoms, the anisotropy of the chemical shift produces fluctuating fields which can cause relaxation.

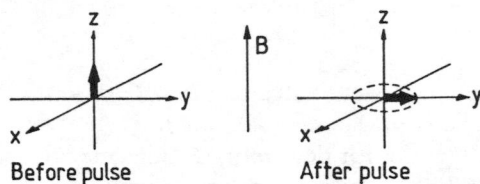

Before pulse   After pulse

Figure 11.14 The effect of a 90° pulse on magnetization

The rate at which relaxation occurs is described by two relaxation times, T_1 and T_2. T_1 is called the spin-lattice relaxation time, and describes the rate at which the magnetization parallel to the magnetic field returns to its equilibrium value after a perturbation. T_2 is called the spin–spin relaxation time and describes the decay of the magnetization in the plane perpendicular to the field. Figure 11.14 shows the total magnetization of a sample in a magnetic field, with the field pointing in the z direction. Slightly more nuclei are oriented with the magnetic field than against it, and so the magnetization has the value M in the z direction and zero in the x and y directions. A rapid radiofrequency pulse is now applied, such that the magnetization is tipped through 90°, and the magnetization now points along the y axis. The effect of the magnetic field is to make the magnetization vector rotate in the xy plane at the Larmor frequency. The magnetization now slowly returns to its equilibrium value, so M_z returns from zero and M_x and M_y return to zero. T_1 describes the rate of the return of M_z to M and T_2 the rate of the return of M_x and M_y to zero.

Any relaxation processes which affect the value of M_z (T_1 processes) alter the total energy of the nuclear spin system, as M_z interacts with the applied magnetic field. These processes must therefore involve exchange of energy between the nuclear spin system and the rest of the sample, the 'lattice'. On the other hand, processes which exchange energy only between pairs of nuclear spins (T_2 processes) do not affect M_z, as the total energy of the spin system remains unchanged, but can reduce the values of M_x and M_y by causing the rotations of individual nuclei to become out of phase, and therefore to cancel each other. For mobile liquids it is often found that $T_2 \sim T_1$, whereas for solids T_2 is typically much less than T_1.

The value of T_2 limits the lifetime of an individual nucleus in its energy state, and therefore, by the uncertainty principle, affects the linewidths of NMR lines. The linewidth at half-height is ($1/T_2 \times \pi$) hertz; for mobile liquids T_2 may be of the order of seconds, producing line broadening of less than 1 Hz, whereas for solids T_2 may be less than 10^{-3} seconds, giving broad bands whose width is measured in kilohertz.

The value of T_1 is of experimental importance as it determines the extent of 'signal saturation'. We have already seen that in NMR there is a very small population difference between the different nuclear spin states; when a radiofrequency field is applied at the Larmor frequency, there is net absorption of energy, which tends to equalize these populations. At the same time, T_1 relaxation processes tend to restore the population difference. If the rate of absorption of energy is high and relaxation is slow, then the population difference decreases rapidly and the energy absorption decreases; the signal is said to be saturated. On the other hand, if the rate of absorption is lower and the relaxation is faster, then the population difference remains constant, as does the signal intensity. Samples with high values of T_1 therefore require the use of low radiofrequency fields, which can produce low sensitivity, whereas samples with low T_1 values can be studied at high radiofrequency power without saturation occurring. In some experiments small quantities of paramagnetic substances are deliberately added to the sample to reduce the value of T_1 and so avoid signal saturation.

A knowledge of relaxation times is important for the measurement and interpretation of NMR spectra. As the understanding of relaxation mechanisms has increased, measurements of relaxation times have become increasingly important. In favourable cases, the values of T_1 and T_2, and the ways in which they vary with parameters such as magnetic field and temperature, can be used to obtain information on the motions of both small and large molecules. Further details are beyond the scope of this book, but can be found in the references given under 'Further Reading'.

I. Experimental NMR methods

A simple form of NMR spectrometer is shown in Figure 11.15. This is a continuous wave (CW) spectrometer, in which the r.f. radiation frequency is held constant and the magnetic field is varied slowly with time, so as to bring all the nuclei to resonance in turn. In practice this is found to be preferable to the alternative of holding the magnetic field constant and varying the radiation frequency.

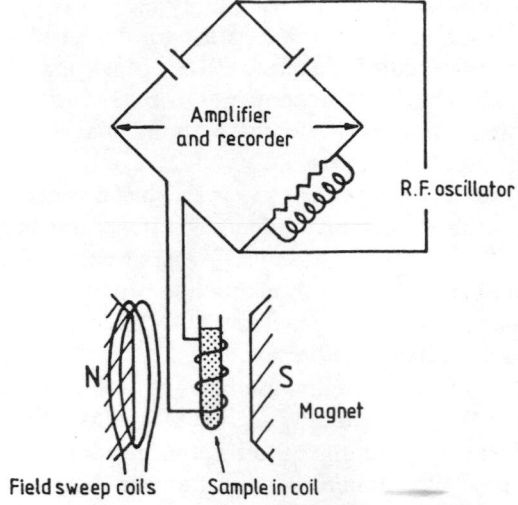

Figure 11.15 The simple form of an NMR spectrometer.

The sample is held in a glass tube, which is placed in a strong magnetic field within an inductance coil at right angles to the field. The coil forms one arm of a balanced radiofrequency bridge, which is fed from a radiofrequency oscillator at the resonant frequency. When the radiofrequency field generated in the coil is in part absorbed by the nuclei, the energy loss causes the bridge to become unbalanced. A signal appears across the detector part of the bridge, and this is amplified and displayed.

Liquid state NMR requires very high resolution to be obtained, of the order of one part in 10^9. This means that the magnetic field must be of high homogeneity, so that each nucleus experiences the same field; this is achieved by placing small additional coils, known as shim coils, on the poles of the magnet. These are adjusted each time the spectrometer is used to correct small field gradients. The homogeneity of the field experienced by the sample is further improved by spinning the sample on its own axis at about 30 Hz, so that all the nuclei experience the same average field.

One of the major problems in NMR is that it is inherently an insensitive technique; the energy absorption depends on the very small population difference of nuclei oriented with and against the magnetic field, and so the sensitivity of a spectrometer can be improved by increasing its magnetic field. The tendency in recent years has been to build magnets of increasing strength, and the most powerful spectrometers now contain what are called superconducting magnets. In fact, it is not the 'magnet' which is superconducting but the coils which produce the field. Fields of about 1.5 T (corresponding to a proton resonance frequency of 60 MHz) have been commonly used in the past, but the most modern machines employ fields up to an order of magnitude larger than this.

Another important experimental development is the advent of Fourier transform (FT) spectrometers. In CW spectrometers the spectrum is scanned slowly, so that each resonance is observed in turn; in FT spectrometers, a strong radiofrequency pulse is applied which stimulates all the nuclei simultaneously. The behaviour of the spins after the pulse is then monitored (the free induction decay, FID) and a computer in the spectrometer applies a Fourier transformation to the signal, which produces the conventional NMR spectrum. In practice, the sample is pulsed repeatedly, often many thousands of times, and the FIDs are added to improve the signal to noise ratio. FT techniques are particularly valuable in samples where the concentration of magnetic nuclei is very low; more complex pulse sequences may also be used to obtain direct measurements of the relaxation times T_1 and T_2.

J. Basic principles of ESR

The electron spin resonance experiment involves the reorientation of the magnetic moment of an electron in a strong field; in many ways it resembles NMR, but it differs from NMR in two important details. Firstly, the Pauli exclusion principle requires that whenever two electrons occupy one orbital their spins should be opposed. It is therefore only possible to reorientate the spin of an electron if the electron is unpaired, and so ESR is restricted to ions and molecules with odd electrons. Closed-shell molecules give no ESR signal. Secondly, because the electron is much lighter than any nucleus, its magnetic moment is much greater, with the result that ESR transitions fall in the microwave region of the spectrum. The energy of an electron in a magnetic field is shown in Figure 11.16: note that the ordering of the states $M_s = +\frac{1}{2}$ and $-\frac{1}{2}$ is the opposite of that for the proton (Figure 11.2) as the electron and proton have opposite charges.

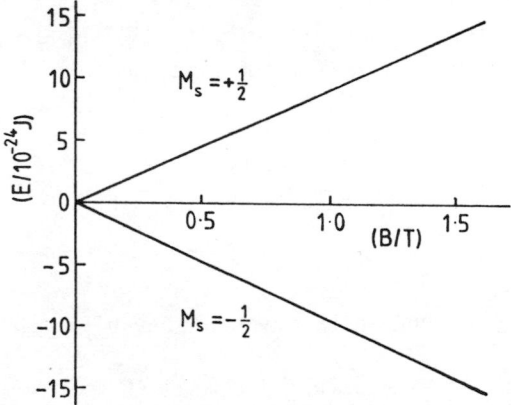

Figure 11.16 Energies of the two states of an electron in a magnetic field

160

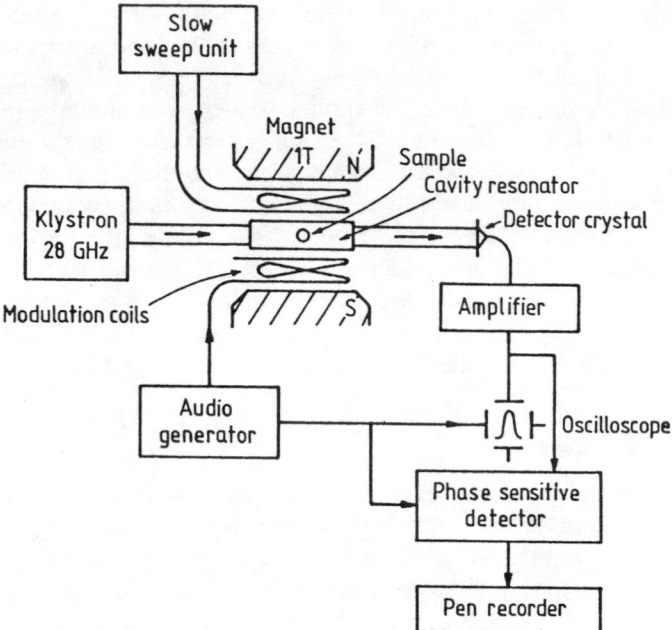

Figure 11.17 Schematic arrangement of an ESR spectrometer

By analogy with NMR we can write

$$h\nu = g\,\mu_B\,B,$$

where μ_B is called the Bohr magneton. The value of μ_B is given by

$$\mu_B = \frac{eh}{2M_e}$$

μ_B has the value of $9.3 \times 10^{-24}\,m^2A$, which is 1837 times greater than μ_N, the nuclear magneton. For a free electron, g has the value 2.0023, so in a field of 1 T,

$$\nu = \frac{2.0023 \times (9.3 \times 10^{-24}) \times 1}{6.626 \times 10^{-34}}\,Hz$$

$$= 2.8 \times 10^{10}\,Hz$$

$$= 28\,GHz.$$

The value of g in molecules may show some deviation from the free electron value.

Figure 11.17 shows schematically a simple ESR spectrometer. The sample may be a liquid or solid, with a volume typically up to 1 cm^3. The microwave radiation is produced by a klystron oscillator and is led to the sample by waveguides, which are hollow rectangular metal pipes. When the sample is

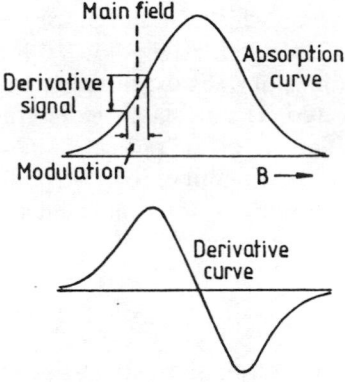

Figure 11.18 Detection of an ESR signal

brought to resonance, radiation falls on a semiconducting detector crystal; the resulting signal is then amplified and displayed.

Because ESR signals are relatively broad, they are measured differently from NMR signals, and this results in ESR spectra having an unfamiliar appearance. A set of coils is added to the sample cell, producing a small extra field which oscillates sinusoidally with time. This is field modulation; it is important that the magnitude of this field is much less than the linewidths of the ESR peaks. The signal from the detector now contains a component which oscillates at the modulation frequency, as shown in Figure 11.18. On the left-hand side of the peak, the absorption varies strongly with time as the gradient of the peak is high. At the centre of the peak, the gradient is now zero and so the amplitude of the absorption is constant. On the right-hand side, there is again strong variation with time, although 180° out of phase with the signal obtained on the left-hand side. The spectrometer now extracts only the component of the signal which oscillates at the modulation frequency, whose appearance is that of the derivative of the original absorption spectrum. Field modulation produces a considerable improvement in the signal to noise ratio, and ESR spectra are normally shown as the derivative curve rather than in the familiar absorption form.

ESR is inherently more sensitive than NMR as the population difference between the spin states is greater. From the Boltzmann law, for an electron resonating at 28 GHz,

$$n_+/n_- = \exp\left(\frac{-h\nu}{kT}\right)$$

$$= \exp -\frac{(6.63 \times 10^{-34}) \times (2.8 \times 10^{10})}{(1.38 \times 10^{-23}) \times 300}$$

$$= 0.9955.$$

Although the populations of the two levels are still almost exactly equal, the difference between them, on which the net absorption of radiation depends, is much greater than in NMR (cf. page 143). ESR spectra can be obtained readily from dilute solutions of radicals; in the most favourable cases, as few as 10^{11} spins can be detected. The probability of a transition is independent of the precise chemical nature of a radical, and so ESR can be used quantitatively to measure the number of radicals in a sample. This is of importance in the assessment of radiation damage and in monitoring free radical reactions.

K. Hyperfine structure

The ESR spectra of many radicals show hyperfine structure, which is caused by the presence of magnetic nuclei in the radical; this structure is often the most important feature of the ESR spectrum. In some ways the hyperfine structure resembles the spin–spin coupling which we saw in NMR, but there are also some important differences.

In liquid samples, where there is rapid molecular tumbling, the coupling between the unpaired electron and a magnetic nucleus is due to the Fermi or contact interaction. The magnitude of this interaction depends on the value of the electron wavefunction at the nucleus in question. As p, d and f orbitals all have nodes at the nucleus on which they are centred, they have zero electron density at the nucleus. They make no contribution to the electron–nuclear coupling. However, s orbitals have a finite electron density at the nucleus and electrons in orbitals with significant s character interact strongly with nuclear magnetic moments.

If an electron interacts with just one nucleus of spin ½, then its ESR signal consists of two peaks of equal intensity, corresponding to the two possible orientations of the nucleus. The separation of the peaks is the coupling constant A, whose value is typically measured in megahertz and which is field-independent. An electron which interacts with two spin ½ nuclei gives an ESR signal with three peaks, with intensities in the ratio of 1:2:1; these correspond to the nuclear orientations ↑ ↑, ↑ ↓ and ↓ ↑, and ↓ ↓. Figure 11.19 shows the ESR spectrum of the CH_3 radical; the electron interacts with three equivalent protons and so four peaks are seen, with intensities in the ratio of 1:3:3:1. If an electron is coupled to a nucleus with spin 1, such as ^{14}N, then the ESR spectrum consists of three lines of equal intensity, corresponding to the nuclear orientations $M_I = -1$, 0 and +1. In many cases an electron couples to more than one set of nuclei, in which case the coupling energies are simply additive. The splitting pattern is obtained by considering the splitting caused by one set of nuclei and then superimposing on each line the splitting pattern from the second set, and so on. Complex splitting patterns can sometimes be clarified by using deuterium substitution; this produces a different splitting pattern, as $I = 1$ for the ^2H nucleus, and also alters the coupling constants, as the magnetic moment of ^2H differs from that of ^1H.

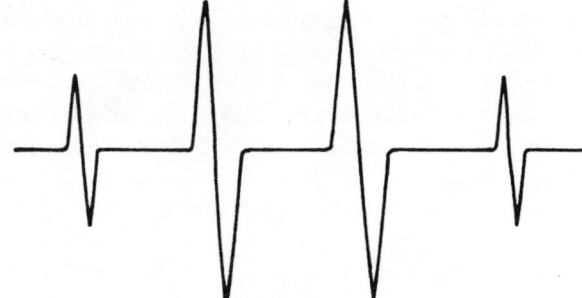

Figure 11.19 The CH_3 spectrum

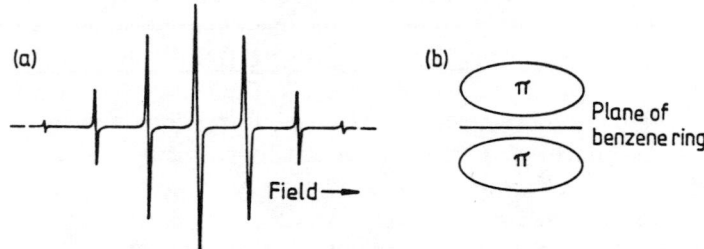

Figure 11.20 (a) The ESR spectrum of the benzene radical anion. (b) Illustration of the node in the π-electron density in the plane of the benzene ring

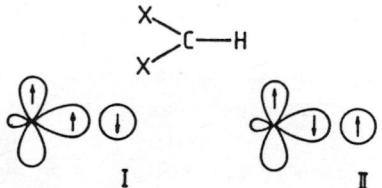

Figure 11.21 The origin of negative hyperfine coupling in C−H fragments: I is favoured over II

It might be thought that in aromatic radicals, where one unpaired electron occupies a delocalized π orbital, that no hyperfine structure would be observable (see Figure 11.20). However, the π electron is found to be influenced by the magnetic moment of the protons in the ring by the mechanisms of 'spin polarization'. Figure 11.21 shows the spin arrangements in a C−H fragment in an aromatic molecule. The presence of an unpaired electron in the p orbital on the carbon atom causes the C−H bond to be spin polarized, with the electron of one spin marginally closer to the C nucleus and the electron of the other spin closer to the H nucleus. The electron closer to the C atom has the same spin as that in the p orbital, as this arrangement has a more favourable exchange energy. The electron closer to the H nucleus therefore has the opposite spin and this electron interacts with the proton magnetic moment. Thus in the radical ion $C_6H_6^-$, the unpaired electron interacts with six equivalent protons and gives a spectrum consisting of seven lines.

One of the most important applications of ESR is the identification of the chemical structures of radicals. The hyperfine splitting can be of great value here as it contains information on the types and numbers of magnetic nuclei and on their environments. ESR also gives information on the electronic structures of radicals. The value of the coupling constant for a free hydrogen atom is 1420 MHz; if the value of the coupling constant of an electron to a hydrogen atom in a radical is A MHz, then we may crudely say that the spin population on that hydrogen atom is $A/1420$, or that the coefficient of the $1s$ orbital based on the atom is the molecular orbital occupied by the unpaired electron and is $(A/1420)^{1/2}$. Similarly, the coupling constant between a $2s$ electron and a ^{13}C nucleus is 3330 MHz; an electron in an sp^3 hybrid orbital would give a value of one-quarter of this, as the orbital has 25 per cent. s character. Coupling constants therefore give information on the hybridization in a radical, and hence on molecular geometry. It is also possible to estimate the degree of covalency in complex ions by observing the coupling of an unpaired electron to a ligand nucleus.

One of the best known applications of coupling constants is the use of the McConnell equation for calculating spin populations in aromatic systems. McConnell suggested that the coupling constant A_H between an unpaired electron and a proton in an aromatic ring was simply proportional to the spin population of the p orbital on the adjacent carbon atom, p_c. Thus

$$A_H = Q \, p_c,$$

where Q is roughly constant for all aromatic radicals. The value of Q can be shown to be -63 MHz by considering benzene, in which the spin population on each carbon atom is one-sixth by symmetry. Spin populations measured in this way may be used to test theoretical methods such as Hückel calculations; agreement in many cases is good.

In solid samples there is a further direct dipolar interaction between the magnetic moments of an electron and nucleus, which is not now averaged to zero by molecular tumbling. However, the magnitude of this interaction is often relatively small compared with the Fermi term, and for a sample where the radicals are not oriented, the overall effect on the spectrum is to produce broader lines. If the radicals are oriented within a crystal, which can occur if a pure solid is subjected to radiation damage, then sharper lines are obtained, and the spectrum changes as the crystal orientation is altered. Detailed information can now be obtained on the structure and environment of radicals.

L. The g value

In NMR we saw that a nucleus in a molecule does not resonate at exactly the same frequency as in a free atom; this difference is the chemical shift and is treated by regarding the field experienced by the nucleus as being different from the external field. In ESR a similar phenomenon is observed, but it is more usual to treat deviations from the free electron case as a property of the electron and not the field. Therefore we write

$$h\nu = g\,\mu_B\,B,$$

where g measures the ratio of the resonant frequency to the applied field. If the value of g deviates from the free-electron value, then the position of the ESR signal in the spectrum shifts to high or low frequency, in a way analogous to the chemical shift in NMR.

The g value of an electron will differ from the free electron value if the applied magnetic field induces orbital motions which produce an extra magnetic field with which the electron spin will interact. In many large organic radicals and first-row transition metal ions this orbital contribution is very small and the g values differ little from the free electron value; thus for NO_2, $g = 1.999$, and for CH_3, $g = 2.00255$. Differences in g values in these cases produce relatively small shifts in ESR spectra; e.g. the difference in resonant frequency at 1 T between CH_3 and NO_2 is about 50 MHz, whereas the total width of the CH_3 hyperfine structure is about 130 MHz. Determination of g values can be of use in identifying free radicals, in the same way as chemical shifts are used in NMR, but in many cases the hyperfine structure is of greater value as it contains more detailed information.

The g values for the later transition metal ions, and for the lanthanides and actinides, show much greater deviations from the free electron value. The situation is further complicated if there is more than one unpaired electron in an ion or if the spin–orbit coupling is very strong. The g value is often strongly anisotropic, and if a solid sample is used, its value in the x, y and z directions may be determined separately. ESR can give very detailed information on the electronic structure of ions with partially filled d and f orbitals, but the interpretation is now complex and each configuration has to be considered separately.

Both NMR and ESR have received tremendous attention in recent years from chemists, physicists and biologists alike. They are topics which require a great deal of study for a full understanding of their applications. Although this chapter is only an introduction, further details can be found in the more advanced texts suggested in the 'Further Reading'.

M. Summary

In NMR a magnetic nucleus is placed in a strong magnetic field and reorientated by light in the radiofrequency region of the spectrum. Many different nuclei can be used—^1H, ^{13}C, ^{19}F, ^{29}Si—but each resonates at its own frequency, and only one nucleus is detected in a single experiment. ^{12}C and ^{16}O do not give signals. The appearance of an NMR spectrum is determined

166

mainly by two factors, the chemical shift and spin–spin coupling. The chemical shift gives information on the immediate environment and the coupling on neighbouring magnetic nuclei. The shape of the spectrum can also be influenced by restricted rotations and chemical reactions, and gives information on the rates of these processes. Solid samples give very broad bands instead of sharp lines; this broadening can be removed by magic-angle spinning.

ESR involves the reorientation of an electron spin in a magnetic field by microwaves; most molecules have closed-shell structures with no resultant electron spin, and ESR is restricted to radicals and ions with unpaired electrons. The frequency of absorption is described by the g value, which may be anisotropic; ESR spectra also show hyperfine structure from interaction of the electron spin with magnetic nuclei. They are used to give information on chemical identity, molecular geometry and details of electronic structure.

N. Problems

1. Calculate the Larmor precession frequency of a ^{13}C nucleus in a field of 1T given that its g-value is 1.405.
2. The diagrams show ^1H spectra of two compounds, both of molecular formula C_3H_6O. The numbers above the peaks give the total numbers of H-atoms represented. Give structural formulae for the compounds shown, and assign the spectra as far as you can.

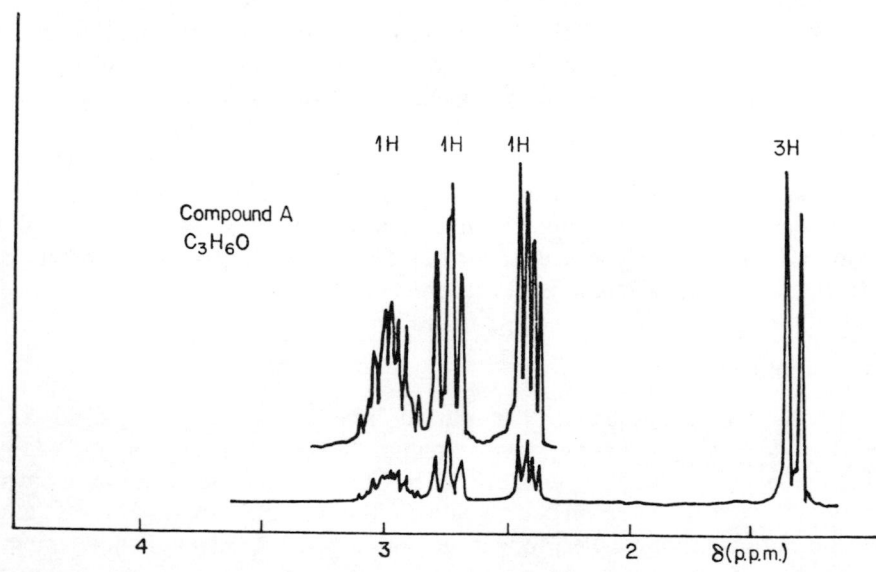

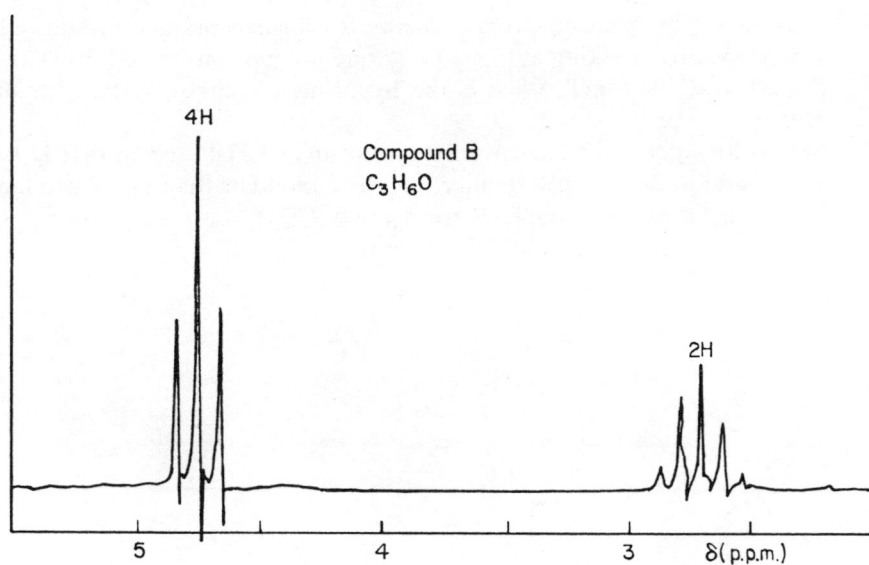

3. The diagram shows the ¹H spectrum of compound C_6H_6ClN. Give a structural formula for the compound, and explain why the peak at $\delta = 3.6$ disappears in D_2O.

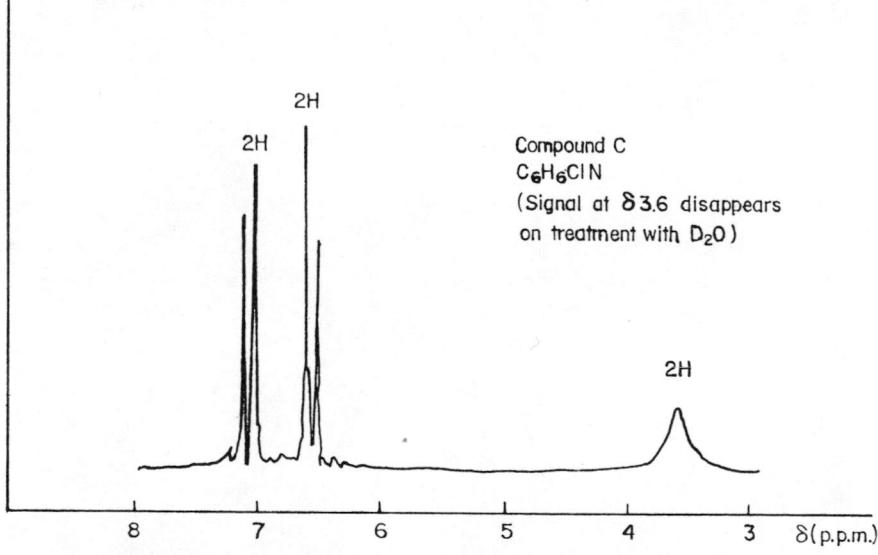

4. For the following molecules state how many proton NMR peaks occur and give the relative intensity of each indicating whether each is a singlet, doublet, triplet, etc.:

$$CH_3CHO, \quad C_6H_6, \quad C_2H_6, \quad CH_3CH_2OCH_2CH_3, \quad CH_3COOCH_3.$$

5. Predict the qualitative form of the ¹⁹F NMR spectrum of $CF_2 = CHF$.

6. Explain why the ESR spectrum of atomic hydrogen consists of two lines. In a spectrometer working at 9.302 GHz one line appears at 357.3 mT and the other at 306.6 mT. What is the hyperfine coupling constant for the atom?

7. Sketch the appearance of the ESR spectrum of CH_2D, given that D has $I = 1$, and a significantly smaller magnetic moment than H. How many lines would there be in the ESR spectrum of CD_3?

Further reading

Atkins, P. W. (1983) *Molecular Quantum Mechanics*, 2nd Edn, Clarendon Press, Oxford.

Banwell, C. N. (1972) *Fundamentals of Molecular Spectroscopy*, 2nd Edn, McGraw-Hill, New York.

Carrington, A. and McLachlan, A. D. (1967) *Introduction to Magnetic Resonance*, Harper and Row, New York.

Coulson, C. A. (1982) *The Shape and Structure of Molecules*, 2nd Edn revised by R. McWeeny, Clarendon Press, Oxford.

Dixon, R. N. (1965) *Spectroscopy and Structure*, Methuen, London.

Gaydon, A. G. (1968) *Dissociation Energies and Spectra of Diatomic Molecules*, Chapman and Hall, London.

Herzberg, G. (1950) *Spectra of Diatomic Molecules*, Van Nostrand, New York.

Herzberg, G. (1945) *Infrared and Raman Spectra of Polyatomic Molecules*, Van Nostrand, New York.

Herzberg, G. (1966) *Electronic Spectra and Electronic Structure of Polyatomic Molecules*, Van Nostrand, New York.

Lynden-Bell, R. and Harris, R. K. (1969) *Nuclear Magnetic Resonance Spectroscopy*, Nelson, London.

McLauchlan, K. A. (1972) *Magnetic Resonance*, Clarendon Press, Oxford.

Murrell, J. N. (1971) *The Theory of Electronic Spectra of Organic Molecules*, Chapman and Hall, London.

Richards, W. G., and Scott, P. R. (1976) *Structure and Spectra of Atoms*, John Wiley, Chichester.

Straughan, B. P., and Walker, S. (1976) *Spectroscopy*, Chapman and Hall, London.

Whiffen, D. H. (1972) *Spectroscopy*, 2nd Edn, Longmans, London.

Fundamental constants and conversion factors

Speed or light c 2.997925×10^8 m sec^{-1}
Boltzmann constant k 1.38066×10^{-23} Jk^{-1}
Planck constant h 6.62618×10^{-34} Js
Avogadro constant L 6.02205×10^{23} mol^{-1}
Vacuum permittivity ε_0 8.854188×10^{-12} J^{-2}C^{-2}M^{-1}
B.hr magneton U_B 9.27408×10^{-24} JT^{-1}
Rydberg constant R 2.179908×10^{-23} J

$$1^e V = 1.602189 \times 10^{-19} \text{ J}$$
$$= 96.486 \text{ kJ mol}^{-1}$$
$$= 8065.5 \text{ cm}^{-1}$$

$$1000 \text{ cm}^{-1} = 1.986 \times 10^{-20} \text{ J}$$
$$= 11.96 \text{ kJ mol}^{-1}$$
$$= 0.1240 \text{ eV}$$

Index